Inorganic Chemistry Concepts
Volume 4

Yoshihiko Saito

Inorganic Molecular Dissymmetry

With 107 Figures

Springer-Verlag
Berlin Heidelberg New York 1979

Professor Dr. Yoshihiko Saito

The Institute for Solid State Physics
The University of Tokyo
Roppongi-7, Minato-ku
Tokyo 106, Japan

ISBN 3-540-09176-9 Springer-Verlag Berlin Heidelberg New York
ISBN 0-387-09176-9 Springer-Verlag New York Heidelberg Berlin

Library of Congress Cataloging in Publication Data. Saito, Yoshihiko, 1920-. Inorganic molecular dissymmetry. (Inorganic chemistry concepts; v. 4) Bibliography: p. Includes index. 1. Molecular theory. 2. Symmetry (Physics) 3. Chemistry, Inorganic. I. Title. II. Series. QD461.S26 541'.2 78-26270

Typesetting: Elsner & Behrens GmbH, 6836 Oftersheim.
Printing and bookbinding: Zechnersche Buchdruckerei, Speyer.
2152/3140-543210

Preface

As early as 1874 van't Hoff and Le Bel introduced the concept of antipodes for molecules containing an asymmetric carbon atom. This was the first insight into the spacial arrangement of atoms in a molecule. These antipodes exhibit opposite optical rotatory power, but it was not possible to determine specific configuration and direction of the rotatory power. The convention of Fischer, however, gained general acceptance. Eighty years later Bijvoet and his co-workers showed that the Fischer convention happens to be in agreement with reality (1951).

Organic stereochemistry is that of tetrahedral carbon atoms, while stereochemistry of co-ordination compounds mainly concerns octahedrally co-ordinated metal atoms. The stereochemistry of octahedrons was founded by Werner. In his magnum opus, Neuere Anschauungen auf dem Gebiete der anorganischen Chemie, are summarised his highlights on optical isomerism, beginning with the third edition published in 1913. After about forty years, the author and his co-workers determined the absolute configuration of the tris(ethylenediamine)cobalt(III) ion. Thus an absolute basis was given to discuss the optical activity and molecular structure of co-ordination compounds.

This book deals with the absolute stereochemistry of transition-metal complexes, the charge-density distribution in them and their circular dichroism spectra. The book is directed to students of inorganic chemistry and to others seeking a general impression of the recent advances in the field.

The basic principles of crystal-structure determination by X-ray and neutron diffraction are briefly described in the hope that the reader may appreciate how the absolute configuration of a dissymmetric molecule can be determined by utilizing anomalous scattering of X-rays or a neutron beam. The procedure of strain-energy minimization of a metal complex is outlined, and the following chapter deals with the isomerism and structures of dissymmetric co-ordination compounds, where octahedral complexes are mainly discussed with reference to their conformational energy. The choice of the complexes is largely determined by the author's interests. In the next chapter the electron-density distribution of transition-metal compounds is described in some detail. This is a rather new field and will play an important role in constructing a theoretical model for optical activity and other chemical and physical properties of the transition-metal complexes. In the last chapter the circular dichroism of transition-metal complexes is discussed where emphasis is laid on tris-bidentate complexes, since they have been most extensively studied.

It is a pleasure to acknowledge the following bodies for permission to reproduce certain figures or parts of figures: Acta Crystallographica, Akademische Verlagsgesellschaft, The Chemical Society, The Chemical Society of Japan, Elsevier Publishing Company and Verlag Chemie.

December, 1978   Yoshihiko Saito

Contents

Chapter I Introduction

1 Preamble

Chemists have not yet completely agreed on a simple definition of co-ordination compounds, because co-ordination compounds differ greatly in nature and in stability. What is meant by a co-ordination compound is perhaps very well laid down in the following definition. A co-ordination compound is a species formed by the association of two or more simpler species capable of independent existence.

For instance, two or more compounds capable of independent existence often combine:

$$AlF_3 + 3NaF \rightarrow Na_3[AlF_6]$$

$$2KCl + PtCl_4 \rightarrow K_2[PtCl_6]$$

These products differ widely in their behaviour, particularly in water. If the crystal structure of these compounds are examined, a grouping of atoms will be recognized in which an atom M is attached to other atoms A or groups of atoms B to a number in excess of the charge or oxidation numbers of the atom M. Such a grouping of atoms is called a complex molecule or a complex ion (or simply complex). For example, crystals of $K_2[PtCl_6]$ consist of potassium ions and the groupings of $[PtCl_6]^{2-}$ in which a platinum atom is surrounded octahedrally by six chlorine atoms. When M is a metal atom the resulting entity is called a metal complex. The atom which is directly attached to the central atom is a co-ordinating atom. A chelate ligand is one using more than one of its co-ordinating (ligating) atoms. If a chelate ligand is co-ordinated to a metal atom, a closed ring is necessarily formed. This is called the chelate ring. The word, chelate was first introduced by Morgan and Drew (1920), which was derived from the Greek χηλή meaning a lobster's claw. For example, the molecule of ethylenediamine, $H_2NCH_2CH_2NH_2$, has two amino groups. The nitrogen atoms are co-ordinating atoms and can form a five-membered chelate ring with a cobalt atom.

The 1913 Nobel Prize in Chemistry was awarded to Alfred Werner for his co-ordination theory. This distinguished work can be summarised as follows. Among many metal complexes the cobalt complexes were of dominant importance for Werner to elucidate the structure of metallic compounds. Due to the inertness of Co(III) and the lability of Co(II), dozens such complexes has already been prepared in simple ways from quite early on. Their distinct colours assisted in their discovery and helped to distinguish different species from one another. Three complexes will be quoted to explain his co-ordination theory.

$CoCl_3 \cdot 6NH_3$ yellow (luteo salt)
$CoCl_3 \cdot 5NH_3$ purple (purpreo salt)
$CoCl_3 \cdot 4NH_3$ green (praseo salt)
                       violet (violeo salt)

For the third salt there are two isomers which can be easily distinguished by their colours. All these salts are soluble in water. If silver nitrate solution is added, all the chloride ions are precipitated for the first compound. In the case of the second salt, only two-thirds of the chloride ion can readily be precipitated and of the rest, one Cl is rather immobile and can only very slowly be precipitated by silver nitrate solution. For the last compound, only one third of the chlorine can readily form silver chloride, and the remaining two Cl's are more tightly bound to the metal atom, by secondary valency (Nebenvalenz). Nowadays, this is called the co-ordinate bond. The nature of the co-ordinate bond as revealed by an accurate determination of the electron-density distribution in crystals will be discussed in Chapter V. If these chemical formulae are rewritten, two kinds of chlorines can be distinguished,

$[Co\,6(NH_3)]Cl_3$
$[Co\,5(NH_3)Cl]Cl_2$
$[Co\,4(NH_3)2Cl]Cl$

There are always six atoms or atomic groups bound to the metal atom by secondary valency. Now what is the spacial arrangement of these six groups? If six identical atomic groups like ammonia enter into combination with the cobalt atom, there are three possibilities: hexagonal planar, trigonal prismatic or octahedral arrangements.
    For the compounds of the type $[M\,^{B2}_{A4}]$, how many isomers are possible? The experimental evidence is that there are only two isomers. This indicates that the six groups are arranged octahedrally around the metal atom.
    For hexagonal planar co-ordination, there are three possible isomers just like o-, m- and p-substitution in benzene derivatives as shown in Fig. 1.2. For trigonal pris-matic co-ordination three isomers are also possible as shown in the figure. In the case of octahedral co-ordination alone, only two isomers are possible. It is to be noted here that at that time chemists had no means of determining the atomic arrangement in

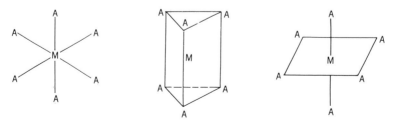

Fig. 1.1. Possible arrangements of six atomic groups around the central metal atom

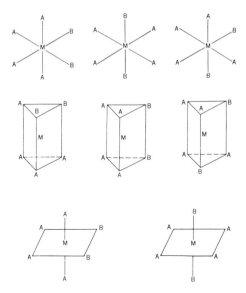

Fig. 1.2. Possible isomers of $[M\,{}^{B_2}_{A_4}]$

solids. Laue's discovery of the diffraction of X-rays by crystals was made in 1912. Werner has come to this conclusion by his ingenious speculation and painstaking experiments. The most conspicuous success of Werner's co-ordination theory was the first isolation of optically active metal complexes in 1911 (Werner and King). The first series of compounds to be resolved were the cis isomers of $[CoAB(en)_2]X_2$ (Fig. 1.3).

In 1912 tris (ethylenediamine)cobalt(III) ions were resolved into optical isomers (Werner). The octahedral arrangement of the ligating atoms was indeed verified in 1922 by the X-ray crystal structure analysis of $K_2[PtCl_6]$ (Scherrer and Stoll) and of $[Co(NH_3)_6]I_3$ in 1926 (Stoll).

This is the main point of Werner's co-ordination theory. His work has been of the greatest importance for the development of co-ordination chemistry. Certainly,

$[CoAB(en)_2]^{2+}$

$[Co(en)_3]^{3+}$

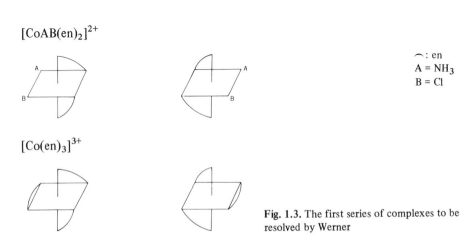

⌒ : en
A = NH_3
B = Cl

Fig. 1.3. The first series of complexes to be resolved by Werner

it was Werner's remarkable ingenuity and powers of intuition that enabled him to deduce these principles from a study of chemical phenomena alone.

However, there remains one unsolved problem. This is the problem of absolute configuration, that is the determination of the exact three-dimensional structure of a particular isomer, so that the position of each atom in the optically active complex is known relative to all other atoms. After the discovery of diffraction of X-rays by crystals in 1912, we had means of determining the atomic arrangement in crystals. Normal X-ray methods, however, do not tell us whether the optically active complex has a particular configuration or one related to this as its mirror image. In other words, it is not possible to assign the absolute stereochemical configurations to enantio-morphically related pairs of the complexes. This is a great challenge which has only recently effectively been met by the advent of the necessary instrumentation and techniques. Forty-two years after Werner's resolution of tris(ethylenediamine)cobalt(III) ions, the absolute configuration of this complex ion was determined by means of X-ray anomalous scattering and the dextro-rotatory isomer, $(+)_{589}[\mathrm{Co(en)_3}]^{3+}$, was found to correspond to the configuration (XIX) on p. 10 (Saito, Nakatsu, Shiro and Kuroya 1955).

Here, we shall briefly recall two general concepts that are basically important in co-ordination chemistry: "symmetry" and "isomerism".

2 Symmetry

The concept of symmetry has been understood for a very long time, and the relationship of the symmetry of an object to its aesthetic appeal has been appreciated from the earliest ages. Nowadays it is of fundamental importance to understand rationally the structure and properties of the complexes. Symmetry is concerned with the relations between the various parts of an object. The units of symmetry are called symmetry elements. Their precise definitions can be best understood in conjunction with their associated symmetry operations. A symmetry operation involves doing something to an object which leaves it in an indistinguishable (not necessarily identical) situation. Thus the existence of a symmetry element can be demonstrated by applying a symmetry operation to a body. The symmetry operations are as follows:

1 Rotation About an Axis of Symmetry C_n

It is convenient to have a shorthand representation for the symmetry operation and element of symmetry for a particular object. Thus a rotation axis is given the symbol C_n, where $(360/n)°$ is the rotation necessary to give an equivalent configuration. Such an axis is said to be an n-fold axis. Hexachloroplatinate(IV) ion, $[\mathrm{PtCl_6}]^{2-}$, for example, has three four-fold axes of rotation through each Cl and Pt atom. It also possesses a three-fold axis through the platinum atom and perpendicular to each face of the octahedron. There are twelve two-fold axes through Pt and each centre of the Cl–Cl edge of the octahedron and through each Pt and Cl atom.

2 Reflection Through a Plane of Symmetry (or a Mirror Plane), m

A plane of symmetry is any plane which divides an object into two equal and similar halves, each of which is a mirror image of the other. Thus the square-planar complex ion, $[PtCl_4]^{2-}$ has a plane of symmetry through four Cl atoms. It also possesses two mirror planes through the Cl–Pt–Cl bond and perpendicular to the plane of $[PtCl_4]^{2-}$ and another two through the midpoint of the Cl–Cl edge and the platinum atom and perpendicular to the co-ordination plane.

3 Inversion Through the Centre of Symmetry (Inversion), i

The complex ion, $[PtCl_4]^{2-}$, for example, is symmetrical about the central metal atom, that is, every point (x, y, z) in the complex will have a corresponding point at $(-x, -y, -z)$. The tetrahedral complex, $[ZnCl_4]^{2-}$ does not possess this type of symmetry.

4 Rotation Followed by Reflection on a Plane Perpendicular to the Axis (Improper Rotation), S_n

This is know as an improper axis of rotation. The symbol S_n is given to the improper axis, where $(360/n)°$ is the angle of rotation as before. A tetrahedral molecule, methane, has S_4.

One can easily see that S_1 and S_2 are equivalent to m (mirror reflection) and i (inversion), respectively.

For further details, the reader may consult textbooks on symmetry groups.

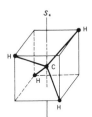

Fig. 1.4. A methane molecule, showing one of the three S_4 axes

3 Isomerism

Isomerism is a rather comprehensive concept embracing several types of structural differences between molecules having the same chemical composition. There are two important isomerisms in co-ordination chemistry; geometrical isomerism and optical isomerism.

Geometrical Isomerism

There are three possible ways of arranging the same six ligands evenly in space around a central metal atom: regular hexagon, trigonal prism and octahedron as illustrated in Fig. 1.

Among these, isomerism in octahedral complexes will be illustrated by considering some general kinds of compounds, since the majority of metal complexes are octahedral.

The compound having the general formula $[Ma_4b_2]$ can exist in two isomeric forms: I and II.

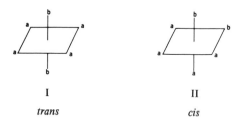

| I | II |
| trans | cis |

In trans isomer I, the two groups in question, b, are opposed to one another about the central metal atom, while they are adjacent for the cis isomers, II. Thus the phenomenon is sometimes called *cis-trans* isomerism. There are also two isomeric forms III and IV for the compound $[Ma_3b_3]$.

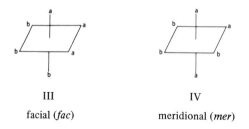

| III | IV |
| facial (*fac*) | meridional (*mer*) |

When two symmetrical bidentate aa replace four a ligands from $[Ma_4b_2]$, two isomers are possible: *trans* V and *cis* VI.

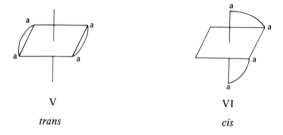

| V | VI |
| trans | cis |

But if the chelate ligand is unsymmetrical, there are five geometrical isomers.

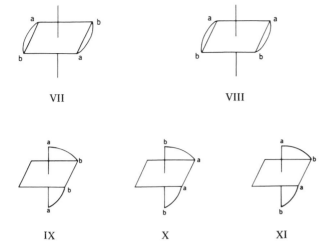

VII   VIII

IX   X   XI

Here, the letters a and b represent different ends of an unsymmetrical bidentate ligand, not necessarily different ligating atoms: propylenediamine, $H_2N-CH(CH_3)-CH_2-NH_2$ is an example.

The tris-chelated complex with symmetrical ligands does not exhibit geometrical isomerism, but two geometric isomers are possible for those involving unsymmetrical ligands.

Isomerism concerning multidentate complexes will be illustrated later with actual examples.

Square Planar Complexes

Two examples of geometrical isomers exhibited by square planar complexes are given below:

$[Ma_2b_2]$

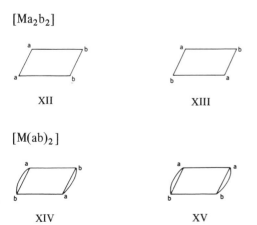

XII   XIII

$[M(ab)_2]$

XIV   XV

Optical Isomerism

Optical isomerism occurs when a complex and its mirror image are not superposable. The most general statement of the criterion for the appearance of optical isomerism is that the complex must not possess an improper rotation axis. The simplest case of chiral molecules are those of the tetrahedral complex, [Mabcd], with symmetry C_1, which implies lack of all symmetry elements (XVI and XVII). These complexes

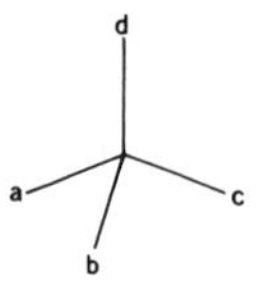

XVI 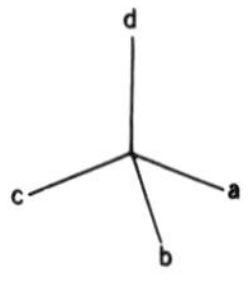   XVII

are chiral and asymmetric, Among transition metal complexes there are many examples of octahedral complexes which are chiral but do not lack all the elements of symmetry. For example, the tris-bidentate complexes such as $[Co(en)_3]^{3+}$ possess D_3 symmetry. They have one three-fold axis of rotation and three two-fold axes which are perpendicular to the former.

Diastereoisomerism

Diastereoisomers are stereoisomers that have the same elements of dissymmetry, some but not all of which are enantiomeric. For example, three isomers of tartaric acid are:

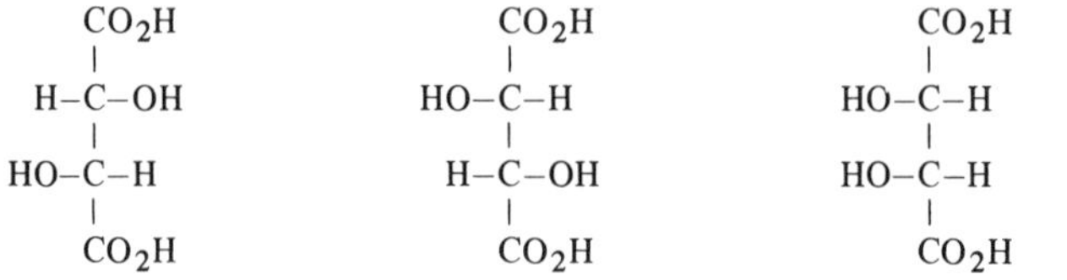

$$
\begin{array}{ccc}
\text{CO}_2\text{H} & \text{CO}_2\text{H} & \text{CO}_2\text{H} \\
| & | & | \\
\text{H}-\text{C}-\text{OH} & \text{HO}-\text{C}-\text{H} & \text{HO}-\text{C}-\text{H} \\
| & | & | \\
\text{HO}-\text{C}-\text{H} & \text{H}-\text{C}-\text{OH} & \text{HO}-\text{C}-\text{H} \\
| & | & | \\
\text{CO}_2\text{H} & \text{CO}_2\text{H} & \text{CO}_2\text{H} \\
R,R & S,S & R,S\text{(meso)}
\end{array}
$$

Here, R,R- and R,S-tartaric acids (or S,S and R,S) are diastereoisomers, whereas R,R and S,S are optical isomers. Another example is Λ-[Co$(-$chxn)$(+$chxn$)_2]^{3+}$ and Λ-[Co$(-$chxn$)_3]^{3+}$ [1].

Conformational Isomerism (Rotational Isomerism)

Conformation means two or more non-identical three-dimensional arrangements of atoms in a molecule that can be interconverted by rotation around one or more single

---

1   Unless otherwise stated all signs of rotation quoted in this book are at the sodium D-lines.

bonds. Isomers due to different conformations are called conformers. In co-ordination chemistry, however, the word "conformation" is often used to describe the spacial distribution of atoms in individual chelate rings and not a whole complex.

4 Designation of the Absolute Configuration

Absolute configuration concerning six-co-ordinated complexes based on the octahedron in this book is designated according to IUPAC proposal (1971, 1970; Thewalt, Jensen and Schäffer, 1972).

Two skew lines that are not orthogonal define a helical system.
By comparing Fig. 1.5 (a) and (c) one may easily see that the two skew lines AA′ and BB′ define a right-handed helix, namely AA′ determines an axis of a helix and BB′ makes up a tangent to the helix and defines its pitch. As far as a qualitative measure of helicity is concerned, the steepness of a helix (pitch) is in general of no importance. Alternately, if BB′ is chosen an as axis of the helix and AA′ as its pitch, a helix of the same hand is obtained. This is because there is a two-fold axis perpendicular to the common normal (in fact there are two such axes and they are perpendicular to each other). Rotation around this axis will bring AA′ to BB′ and vice versa. This means that the second choice will give rise to the same helicity as the first one. In the same way Fig. 1.5 (b) defines a left-handed helix. The Greek letter delta (Δ referring to configuration, δ to conformation) is associated with the two skew lines shown in Fig. 1.5 (a) and the Greek letter lambda (Λ for configura-

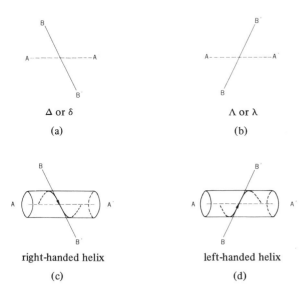

Δ or δ

(a)

Λ or λ

(b)

right-handed helix

(c)

left-handed helix

(d)

Fig. 1.5. Pairs of non-orthogonal skew lines are shown in projection along the common normal. The full line BB′ is above the plane of the paper, the dotted line AA′ below the plane. Lower figures **(c)** and **(d)** show corresponding helical systems to **(a)** and **(b)** respectively

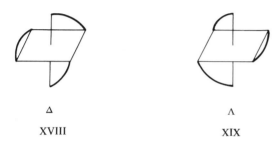

Δ   Λ
XVIII   XIX

tion, λ for conformation). The absolute configurations of the two optical isomers
of a tris-bidentate complex with symmetrical ligands are shown above (XVIII and
XIV), where any pair of two edges of an octahedron on which a chelate ring is spanned defines two skew lines Δ or Λ.

5 Abbreviation of the Ligands

The following abbreviation of the ligands are used throughout this book.

acac	acetylacetonate
ala	alaninate
asp	aspartate
chxn	*trans*-1,2-diaminocyclohexane
	(*trans*-1,2-cyclohexanediamine)
cptn	*trans*-1,2-diaminocyclopentane
	(*trans*-1,2-diaminocyclopentane)
dien	diethylenetriamine
edda	ethylenediamine-N,N′-diacetate
eddda	ethylenediamine-N,N′-diacetate-N,N′-di-3-propionate
edta	ethylenediamine-N,N,N′,N′-tetraacetate
en	ethylenediamine
glut	glutamate
gly	glycinate
linpen	1,14-diamino-3,6,9,12-tetraazatetradecane
	(linear pentaethylenehexamine)
mal	malonate
N-Me-ala	N-methylalaninate
N-meen	N-methylethylenediamine
mepenten	N,N,N′,N′-tetrakis(2′-aminoethyl)-1,2-diaminopropane
MeTACN	2-methyl-1,4,7-triazacyclononane
ox	oxalate
penten	N,N,N′,N′-tetrakis-(2′-aminoethyl)-1,2-diaminoethane
pn	propylenediamine
	1,2-diaminopropane

pro	prolinate
ptn	2,4-diaminopentane
sar	sarcosinate
sarmp	sarcosinate-N-monopropionate
sep	1,3,6,8,10,13,16,19-octaazabicyclo[6.6.6]eicosane
tame	1,1,1-tris(aminoethyl)ethane
R-2,2′,2-tet	5(*R*)-methyl-1,4,7,10-tetraazadecane
3(*S*)8(*S*)-2′,2,2′-tet	3(*S*),8(*S*)-dimethyl-1,4,7,10-tetraazadecane
R,R-2,3″,2-tet	5(*R*),7(*R*)-dimethyl-1,4,8,11-tetraazaundecane
N,N-Me₂-*R,S*-2,3″,2-tet	6(*R*),8(*S*)-dimethyl-2,5,9,12-tetraazatridecnae
R,R,R,R-3″,2,3″-tet	1(*R*),3(*R*),8(*R*),10(*R*)-tetramethyl-4,7-diazadecane-1,10-diamine
tetraen	1,11-diamino-3,6,9-triazaundecane (tetraethylenepentamine)
thiox	thiooxalate
tmd	1,4-diaminobutane
tn	trimethylenediamine 1,3-diaminopropane
trdta	trimethylenediamine-N,N,N′,N′-tetraacetate
TRI	tribenzo[b,f,j]-[1,5,9]triazacyclododecahexaene
trien	triethylenetetramine N,N′-bis(2′-aminoethyl)-1,2-diaminoethane (2,2,2-tet)

Chapter II X-Ray Diffraction

1 Introduction

In this chapter an outline of crystal structure analysis will be described. It is not the purpose of this chapter to give the details of the method but to explain a basic idea of X-ray diffraction.[2] Emphasis will be laid on anomalous scattering of X-rays and its application for the determination of absolute configuration of dissymmetric molecules.

The physical method used for elucidating the structures of complexes falls into two classes: those that yield detailed information about the whole structure of the molecule and those that yield fragmentary information concerning individual bonds or particular groups of atoms in a molecule. The first class includes X-ray, neutron and electron diffraction, the second, various kinds of spectroscopy of the region ranging from microwave to ultraviolet, the measurement of electric and magnetic moments, optical rotatory dispersion and circular dichroism.

Although chemical methods of investigation leave no doubt about the general structural features of metal complexes, such as the octahedral co-ordination and the existence of the chelate rings, X-ray crystal structure analysis reveals the lengths of chemical bonds, the angles between them and electron-density distribution that could not be gained by purely chemical methods. Moreover, the distance between the atoms of neighbouring molecules and ions can be measured with the same precision as those between the atoms in the molecule itself. The quantitative knowledge of intermolecular distances gained in this way is of the greatest importance, especially in structures involving special interactions such as hydrogen bonding or direct metal-metal interactions. It also gives an important clue to the discussion of the mode of association of the complex ion with other ions in solution.

When the X-ray method was first applied to co-ordination compounds, the analysis was often taken no further than a determination of the general shape and symmetry of the complexes. Nowadays, with the help of greatly improved techniques including the use of electronic computers and automated single-crystal diffractometers, the method is now developing towards two extremes: one is the determination of the structure of large molecules and the other is accurate determination of

---

2   Those readers who are interested in the detailed procedures of crystal structure analysis may refer to a standard textbook, for example: (Woolfson, 1970).

electron-density distribution in comparatively simple molecules. For instance, structures as complex as that of vitamin B_{12}, a cobalt(III) complex consisting of 105 atoms, have been successfully solved. Moreover, the structures of complicated protein molecules such as haemoglobin (this is also a complex containing Fe atoms and with about 5,000 atoms) have been determined. These structures are the most complicated that have yet been solved although not in such complete detail as vitamin B_{12}, and they have added considerably to the knowledge of the pattern of living matter.

When the method is carried to its other extreme, it is possible to reveal the accurate distribution of the bonding electrons in a molecule of sufficient complexity to be of chemical interest. This topic will be discussed in Chapter V. It is possible to determine the absolute configuration of a particular enantiomer by the use of anomalous scattering of X-rays. The knowledge of absolute configuration has made a great contribution to the understanding of molecular dissymmetry. This subject will be covered in Chapters IV and VI.

2 Space Lattice

A crystal consists of a large number of three-dimensional repetitions of a basic pattern of atoms. Fig. 2.1 shows an example of a crystal structure, that of potassium hexachloroplatinate (IV), $K_2[PtCl_6]$. It consists of potassium ions and octahedral hexachloroplatinate (IV) ions. If the centre of each platinum atom is joined to those of its neighbours, the structure is divided into a number of identical units of pattern, each of which is a cube in this particular case. This is called the unit cell. The shape and dimensions of the unit cell, that is the three different sorts of edges and the angles between them, are characteristic for each crystal species. These are called cell constants. The directions of the three edges are crystal axes. If each repeating unit of pattern is represented by a point, there arises a three-dimensional

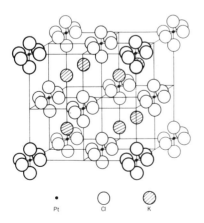

Pt   Cl   K

Fig. 2.1. Arrangement of ions in $K_2[PtCl_6]$

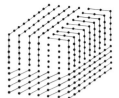

Fig. 2.2. A space lattice, showing its division into a set of parallel equidistant lattice planes

periodic arrangement of points. This is the space lattice. Fig. 2.2 shows a space lattice. A characteristic feature of the space lattice is that it can be divided into a set of parallel equidistant planes on any one of which all the lattice points lie. Such planes are called lattice planes. A few possible ways of dividing the space lattice into lattice planes are illustrated in Fig. 2.2. Each set of these parallel planes is characterised by three integers such as (110), (321) etc., they are called Miller indices of the plane. This system of characterisation uses the three integers that are the reciprocals of the intercepts the plane makes with the three crystal axes. The idea is most easily understood for the cubic system. The plane shaded in Fig. 2.3 (a) is a (100) plane, the intercepts made on the a, b and c axes by the plane being 1, ∞ and ∞, respectively. Hence the Miller indices are $1/1$, $1/\infty$, $1/\infty$. In the same way the plane shaded in Fig. 2.3 (b) is (110) and that in (c) is (321). If any intercept is made on the negative direction, the corresponding index becomes negative, as shown in Fig. 2,3 (d).

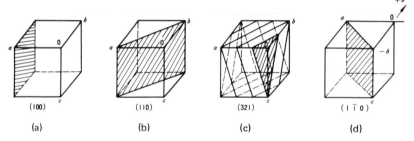

(100)	(110)	(321)	$(1\bar{1}0)$
(a)	(b)	(c)	(d)

Fig. 2.3. Miller indices of lattice planes

3 Symmetry of Crystal Lattices: Space Groups

A solid body or a geometrical figure can be constructed to display any desired symmetry. Order of rotation axis may range from 1 to ∞. These rotation axes may be combined with a mirror plane or with an inversion. The symmetry operations applicable to crystals are, however, strictly limited. For instance, a crystal having a five-fold rotation axis has never been observed. A crystal formed by twelve pentagonal facets does exist, but it has no five-fold symmetry. The distinct rotation axes applicable to crystals are, in fact, only 1, 2, 3, 4, and 6. This is the direct consequence of the lattice structure of crystals. To show this practically, take about 20 pentagonal cardboards of the same size cut out from a large piece and try to pack them tighly on a table. You will find it impossible to do so: there are always some spaces left. The only figures that can be used are the parallelogram (two-fold symmetry), the equilateral triangle (three-fold symmetry), the square (four-fold) and the regular hexagon (six-fold). Any other figure cannot be packed closely without leaving any gaps. To sum up, all the distinct symmetry operations applicable to crystals including the proper and improper rotation axes ar limited to eight as follows: 1, 2, 3, 4 and 6; $\bar{1}(i)$, $m(\bar{2})$ and $\bar{4}$ (S_4).

Bravais (1850) investigated all possible combination of these symmetry elements in lattices and showed that 14 and only 14 different tpyes of lattice belonging to seven crystal systems, can be distinguished. An actual crystal structure like that of $K_2[PtCl_6]$ (Fig. 2.1) may be regarded as an infinitely repeating pattern based on the lattice principle, extended in all directions. The problem is how many distinct combinations of symmetry elements are possible that are applicable to the atomic arrangements in crystals. The eight symmetry elements mentioned above are not enough to describe the symmetry of crystal structures. The symmetry operations so far considered have always been such as to leave at least one point of the body unmoved. The space lattice can be brought into self-coincidence in a new way, namely by translations along any of the lattice directions. Thus the new symmetry operations are obtained by combining a rotation axis with a translation and by combining a plane of symmetry with a translation: the resulting symmetry operations are called screw axis and glide plane, respectively. Fedrow (1885), Schoenflies (1891) and Barlow (1895) analysed carefully the combination of these symmetry elements and showed that there are 230 essentially different ways of combining them. The collection of symmetry operations present in any given crystal structure forms a self-consistent set, or a group in the mathematical sense. Thus they are called 230 space groups. They are all listed and described in volume I of the International Tables for X-ray Crystallography (1959).

4 Diffraction of X-Rays, Bragg's Equation

When the incident beam of X-rays passes over the atoms in a crystal, each atom scatters the incident X-rays. Owing to the periodic arrangement of atoms constructive interference between scattered wavelets takes place to give diffracted beams. Let us consider a very simple structure consisting of one atom on each unit of pattern on a space lattice. If such a three-dimensional point array is to produce a diffracted beam, the diffracted waves from all the atoms must be in phase. Waves scattered from some atoms have the same path length; those from the other atoms will have one wavelength behind the first set and so on. This condition is given by Bragg's law. Figure 2.4 is an edge-on view of successive lattice planes with Miller index (hkl), P_1P_1', P_2P_2', P_3P_3' ... with an interplanar spacing of $d(hkl)$. Diffraction takes place as if the in-

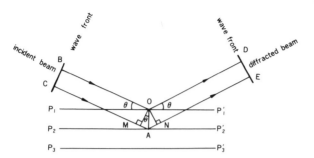

Fig. 2.4. Bragg's condition

cident X-ray beam were reflected by the lattice plane. One can easily see that if the angle of incidence, say BOP_1 is equal to the angle of reflection DOP_1', the path length is the same for all the atoms on P_1P_1', irrespective of the value of θ. However, for waves from P_1P_1' and that from P_2P_2' to give constructive interference, the value of θ is strictly restricted by the requirement that the path difference, MA + AN must be an integral multiple of the wavelength λ (Fig. 2.4).

Therefore,

$$MA + AN = 2d(hkl) \sin \theta, \qquad 2d(hkl) \sin \theta = \lambda \tag{2.1}$$

This is Bragg's equation. If this condition is once fulfilled, the reflected waves from successive lattice planes are all in phase and give strong diffracted beams owing to the periodicity of the lattice. Since λ is known, the interplanar spacing $d(hkl)$ can be calculated by means of Eq. (2.1), if the direction of the diffracted beam is known. The unit cell dimensions can be derived from the values of $d(hkl)$ and the geometrical relations between the diffracted beams.

5 The Geometrical Structure Factor

An actual crystal is not a simple array of points. Each pattern unit generally consists of a group of atoms. Each atom is not a point: electrons which scatter X-rays are distributed around the nucleus like a cloud within a range comparable with interatomic distances. We can show that the form of pattern unit affects the intensities of diffracted beams but it does not affect their positions (direction of the diffracted beam).

If the repeating pattern consists of two kinds of atoms 1 and 2 (Fig. 2.5), we may choose one of these P_1 arbitrarily as a lattice point and as an origin for co-ordinates. Draw in the traces of a lattice plane (hkl) through this atom and one representing the next lattice point P_1'. Inspection of the figure at once reveals that the resulting structure is a superposition of two space lattices of the same dimensions, one consisting of atoms 1 and the other consisting of atoms 2. Both lattices are parallel

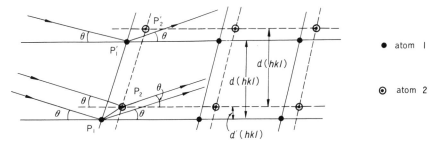

Fig. 2.5. Diffraction of X-ray waves from a lattice composed of two kinds of atoms, 1 and 2

and shifted by P_1P_2. If Bragg's Eq. (2.1) holds, all the waves scattered from each lattice are in phase. Thus the condition for constructive interference is the same for both lattices. The wave diffracted from lattice 2 is, however, not in phase with that diffracted from lattice 1 and the resultant intensity of the diffracted beam from the whole lattice will be affected owing to the phase shift corresponding to P_1P_2. The path difference of the waves scattered by P_1 and P_1' is $2d(hkl)\sin\theta = \lambda$ (for the first order reflexion). In angular measure, the phase difference between the two waves is 2π and the result is complete reinforcement to give a wave of double the amplitude. For the wave scattered in this direction from the atom 2 at P_2, situated at a distance $d'(hkl)$ across the plane, the path difference compared to the wave from the standard atom P_1 is $2\pi d'(hkl)/d(hkl)$ in angular measure. As shown in Appendix II-1, this phase difference is given by $2\pi(hx_2 + ky_2 + lz_2)$, where x_2, y_2, and z_2 are the fractional coordinates of atom 2. The two waves are characterised by the amplitudes f_1 and f_2 in terms of the scattering amplitude of a free electron under the same experimental condition, which depend upon the scattering power of the atom for this angle of reflexion[3] and the phase difference between the two waves is given as mentioned above. The reusltant amplitude $|F(hkl)|$ is obtained by combining the waves scattered by lattices 1 and 2. This can easily be done by adding the vectors representing the component waves, as in Fig. 2.6.

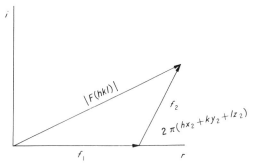

Fig. 2.6. Combination of the two waves of different amplitudes and phase angles

---

3   When X-rays are scattered by atomic electrons, destructive interference between the wavelets scattered from various parts of the electron cloud takes place to an increasing extent as θ increases: $f(hkl)$ therefore falls off from its initial value, $f = Z$, with increasing θ (Z: atomic number).

Expressed analytically

$$F(hkl) = f_1(hkl) + f_2(hkl)\, e^{2\pi i(hx_2 + ky_2 + lz_2)} \tag{2.2}$$

If the repeating pattern consists of N atoms (in other words, there are N atoms in the unit cell), an extension of the treatment which led to (2.2) gives the general expression for $F(hkl)$ as follows:

$$F(hkl) = \sum_{j=1}^{N} f_j(hkl)\, e^{2\pi i(hx_j + ky_j + lz_j)} \tag{2.3}$$

This is the most important equation in crystal structure determination. $F(hkl)$ is known as the geometrical structure factor and $f_j(hkl)$ is the atomic scattering factor. The structure factor is characterized by an amplitude $|F(hkl)|$ and a phase angle $\alpha(hkl)$. They can be evaluated as follows:

$$|F(hkl)| = \sqrt{A^2(hkl) + B^2(hkl)} \tag{2.4}$$

$$\alpha(hkl) = \tan^{-1} B(hkl)/A(hkl) \tag{2.5}$$

where

$$A(hkl) = \sum_{j=1}^{N} f_j(hkl)\, \cos 2\pi(hx_j + ky_j + lz_j)$$

$$\tag{2.6}$$

$$B(hkl) = \sum_{j=1}^{N} f_j(hkl)\, \sin 2\pi(hx_j + ky_j + lz_j)$$

The summations must be taken over all the atoms in the unit cell. Some of these atoms are, however, related to others by a symmetry element of the space groups: therefore (2.3) can be simplified to some extent. If an approximate trial structure can be derived, for instance, from known molecular geometry, packing consideration and other clues to the arrangement of molecules, such as anisotropy of optical properties of single crystals, combined with a knowledge of the space groups, the values of $F(hkl)$ can be calculated from (2.3) and compared with the observed magnitudes of $F(hkl)$. If the agreement is not satisfactory, the co-ordinates are adjusted to give better agreement. The final stage of this refinement is usually carried out on a computer by using the least-squares method. The estimate of agreement is usually given by the R factor:

$$R = \frac{\sum \big| |Fo(hkl)| - |Fc(hkl)| \big|}{\sum |Fo(hkl)|}$$

where $Fo(hkl)$ and $Fc(hkl)$ are the observed and calculated structure factors. Most of the refined structures have R value less than 0.10. In an accurate determination the R value reduces to less than 0.050.

6 Electron-Density

Instead of finding the atomic positions by trial and error, a more direct approach to the problem can be made by the use of Fourier synthesis. The expression for the structure factor given in Eq. (2.3) is obtained based on the assumption that all the atomic electrons are concentrated into a number of spherically symmetrical atoms located at points (x_j, y_j, z_j). This is not a good approximation. In an actual crystal the electrons are generally distributed aspherically between the atomic nuclei owing to bond formation and the electron-density varies continuously from point to point. Thus it is desirable to give a more general definition of the structure factors, which are also necessary to develop a method of structure determination based on the use of Fourier series.

Let $\rho\,(xyz)$ be a distribution function of the electron-density within the unit cell. This density will be expressed in electronic units, so that $\rho(xyz)\,Vdxdydz$ will give the number of electrons in the volume element $Vdxdydz$, where V is the volume of the unit cell. The electrons in each volume element will contribute to the resultant amplitude from the whole unit cell content and, for hkl reflexion, the phase change due to path difference with respect to the origin at the point x, y, z is $2\pi(hx + ky + lz)$. The resultant vector is therefore

$$F(hkl) = V \int_0^1 \int_0^1 \int_0^1 \rho(xyz)e^{2\pi i(hx+ky+lz)}dxdydz \qquad (2.7)$$

$\rho(xyz)$ varies periodically from one unit cell to the other throughout the crystal, since the crystal is a periodic three-dimensional array of atoms. Because of this periodicity $\rho(xyz)$ can be represented by means of a Fourier series in the general form,

$$\rho(xyz) = \sum_{h'}\sum_{k'}\sum_{l'}^{\infty} c(h'k'l')e^{-2\pi i(h'x+k'y+l'z)} \qquad (2.8)$$

h', k' and l' being integers and $c(h'k'l')$ is at present the unknown coefficient. We substitute this series (2.8) in the general expression for the structure factor (2.7) and obtain

$$F(hkl) = \int_0^1 \int_0^1 \int_0^1 V\rho(xyz)e^{2\pi i(hx+ky+lz)}dxdydz$$

$$= V \int_0^1 \int_0^1 \int_0^1 \sum_{h'}\sum_{k'}\sum_{l'} c(h'k'l')e^{2\pi i\{(h-h')x+(k-k')y+(l-l')z\}}dxdydz \qquad (2.9)$$

On integrating, every term is zero except that with $h = h'$, $k = k'$ and $l = l'$, which gives

$$F(hkl) = V \int_0^1 \int_0^1 \int_0^1 c(h'k'l')dxdydz \qquad (2.10)$$
$$= Vc(hkl)$$

$$c(hkl) = F(hkl)/V \tag{2.11}$$

Thus the distribution function of the electron-density $\rho(xyz)$ can be expanded as a three-dimensional Fourier series whose coefficients are the structure factors:

$$\rho(xyz) = 1/V \sum_{h}^{\infty} \sum_{k} \sum_{l} F(hkl)e^{-2\pi i(hx+ky+lz)} \tag{2.12}$$

By using the relations

$$F(hkl) = A(hkl) + i B(hkl) \tag{2.13}$$

and

$$\tan \alpha = B(hkl)/A(hkl) \tag{2.14}$$

Eq. (2.12) can be rewritten in the form

$$\rho(xyz) = 1/V \sum_{-\infty}^{\infty}\sum\sum |F(hkl)| \cos \{2\pi(hx + ky + lz) - \alpha(hkl)\} \tag{2.15}$$

If the electron-density distribution $\rho(xyz)$ could be evaluated over a sufficient number of terms, the result would provide a complete solution of the crystal structure. Unfortunately, the matter is not so straightforward, since the magnitude of $F(hkl)$ can be evaluated easily from the measured intensity of the diffracted beam but the phase angle $\alpha(hkl)$ cannot be directly measured. This is the well-known "phase problem" of X-ray analysis. Inability to determine the phase of the reflected radiation experimentally forces the crystallographer to use a variety of more or less indirect methods. The most important of these methods is known as the heavy atom method, which makes use of an atom of considerable scattering power in the structure. In co-ordination compounds the central metal atom acts as a heavy atom. The position of the heavy atom can be deduced from Patterson synthesis.

The Patterson function is given by a series:

$$P(uvw) = 1/V \sum_{-\infty}^{\infty}\sum\sum |F(hkl)|^2 \cos 2\pi(hu + kv + lw) \tag{2.16}$$

This is a Fourier series with the squares of the structure amplitudes as coefficients and does not require a knowledge of phases. It can be shown that the peaks in $P(uvw)$ represent interatomic distances: if there is a peak at (u, v, w) in the Patterson synthesis it means that there are atoms whose co-ordinates differ by these values; and the peak heights are proportional to the product of the atomic numbers of the atoms concerned. If there are a few atoms (at least two) of a higher atomic number than the rest of the atoms in the unit cell, then the peaks due to these heavy atoms may well stand out in the majority of minor peaks and give clear interatomic vectors. With the aid of these interatomic vectors between the heavy atoms, combined with a knowledge of the space group, we can easily locate the heavy atoms in the unit cell.

Approximate phase angles can then be calculated on the basis of heavy atom positions. A preliminary Fourier synthesis can be carried out by the observed structure amplitudes and the calculated phase angles. The electron-density map so obtained indicates heavy atom positions as expected and may reveal approximate positions of lighter atoms which were not taken into account in the calculation leading to the assignment of phases. The new positions are used to recalculate the phase constants and a second synthesis is carried out. The process is repeated until the phase angles remain the same after recalculation. This procedure consitutes a direct method of adjusting the atomic co-ordinates towards more probable values. It is known as Fourier refinement. At the final stage, the atomic co-ordinates are further refined by the least-squares method described earlier. The atomic co-ordinates are adjusted to minimize $\Sigma\, w(\,|Fo(hkl)| - |Fc(hkl)|)^2$, w being an appropriate weighting function.

7 Thermal Vibration

Atoms in crystals vibrate at ordinary temperature with frequencies very much lower than those of X-rays. At any one instant some atoms are displaced from their mean positions in one direction: consequently, diffracted X-rays would not be exactly in phase and the intensity of the diffracted beam is thus reduced more than would be the case if all the atoms are at rest. Observed electron-distribution is the time average of such random displacements of atoms. The ratio between the actual intensity of a diffracted beam and the intensity which would occur if there were no thermal vibrations is $\exp - [2B(\sin\theta/\lambda)^2]$, where B is a constant. Thus $f_j(hkl)$ in Eq. (2.3) must be replaced by $f_j \exp - [2B(\sin\theta/\lambda)^2]$. This expression was derived by assuming that all the atoms vibrate with equal amplitude and the vibration of the atoms have the same magnitude in all directions (isotropic thermal vibration). This is not strictly true: the amplitude of vibration must be different for different kinds of atoms and some atoms may exhibit marked anisotropic vibrations. The expression for anisotropic vibration is more complicated: f_j must be multiplied by

$$\exp\left[-(\beta_{11}h^2 + \beta_{22}k^2 + \beta_{33}l^2 + 2\beta_{12}hk + 2\beta_{23}kl + 2\beta_{13}hl)\right].$$

Values of B or β_{ij}'s can be estimated from the observed structure amplitudes by the least-squares method. Anisotropic motion of a particular atom in crystals can be conveniently expressed as ellipsoids of thermal motion which are easily calculated on the basis of β_{ij}'s.

8 Experimental Procedure

A source of homogeneous X-rays of known wavelength is used for crystal structure determination. A single crystal specimen bathed in the monochromatic beam will not in general be in a position to produce any diffracted beams. But if it is slowly rotated around some crystal axis, one lattice plane after another will come to a reflecting position, satisfying Bragg's condition, and the diffracted beam will flash out. The diffracted beam will be recorded on a photographic film surrounding the crystal or in the scintillation counter set at the expected position. It is necessary to collect as many reflexions as possible from different lattice planes, in order to calculate the electron density by Eq. (2.15). Photographically, the intensity data are most widely collected by the Weissenberg method. In the Weissenberg camera, a crystal specimen is mounted on a goniometer head and is oscillated to and fro around some crystal axis. The crystal is completely bathed in a narrow beam of monochromatic X-rays. Diffracted beams will be recorded on a film which is inserted into a co-axial cylindrical film holder. The film holder can move along the direction of its axis and it is mechanically coupled to the rotation of the crystal in such a way that it slides forwards and backwards as the crystal oscillates. Between the crystal and X-ray film is a cylindrical screen which is also co-axial with the rotation axis of the crystal. The screen has a circular slit and this slit can be so adjusted that only a certain groups of reflexions passes through it. In this way, one co-ordinate on the photograph gives the angular setting of the crystal and the other gives the spacing for each reflexion. Thus the assignment of the indices of reflexions is quite straightforward. The geometrical relations between any two planes are also known. The unit cell dimensions can be calculated from the co-ordinates of the recorded reflexions. The intensities of reflexions are estimated visually or measured by means of a microdensitometer.

Nowadays automatic X-ray diffractometers are becoming standard intruments for collecting intensity data in an increasing number of laboratories. The most accurate measurements of single crystal reflexions are made in this way. Fig. 2.7 is a schematic drawing of a typical fully automatic four-circle diffractometer most widely used. The term "four-circle" refers to the number of rotational motions available: three of these, ϕ, χ and ω, are associated with the crystal and one, 2θ with the counter. The ϕ circle carries a goniometer head supporting the crystal, which is mounted on the χ-circle. The χ-circle is carried on the ω-circle which is co-axial with the 2θ-circle. The χ-axis is normal to the ω-axis. The counter rotates about a vertical 2θ-axis so that the plane containing the incident and diffracted beam is always horizontal. For a particular reflexion the counter is set at a position corresponding to

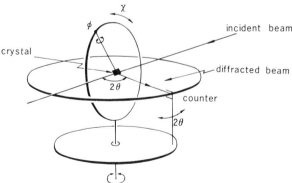

Fig. 2.7. Four-circle diffractometer

the Bragg angle. Once this has been done, the diffracted beam can be produced by positioning the crystal appropriately by rotating around the three axes, ϕ, χ and ω. Once the setting of the crystal is known in terms of the ϕ, χ and ω scale reading for a number of reflexions (at least two), then the three setting angles corresponding to other reflexions can be calculated on the basis of lattice dimensions. A computer built in or linked to the diffractometer, can supervise the process of collecting a complete set of diffraction data without any intervention by the crystallographer. The measured intensities are usually punched out on cards or paper tape, which can be used as an input for further calculations.

9 Process of Crystal Structure Determination

The determination of a crystal structure normally proceeds in three distinct stages. The first is the determination of unit cell dimensions and the space group. The second stage is measurement of the intensities of the Bragg reflexions and calculation from them of structure amplitudes corrected for various geometrical and physical factors depending on the experimental conditions. The third is the determination of atomic co-ordinates. This stage includes the solution of the phase problem, deduction of approximate atomic positions and finally their refinement so as to obtain the best possible agreement between the $Fo(hkl)$ and $Fc(hkl)$'s. The first and the second stages are largely a matter of routine. Approximate unit cell dimensions can be derived from the measurement of co-ordinates of diffraction spots recorded on Weissenberg photographs and refined on the basis of diffractometer data. From the volume of the unit cell and the density of the crystal the number of formula units, Z, in the unit cell can be calculated

$$Z = \frac{\rho V \times 0.6023 \times 10^{24}}{M} \tag{2.17}$$

where ρ is the density, V the volume of the unit cell and M the formula weight. Information about the space group is accessible from a simple study of reflexions recorded on Weissenberg photographs, the systematically absent ones being noted. Some of the space groups can be determined uniquely from such systematic absences. When ambiguity remains, rigorous determination of the space group can be achieved first by the success of the structure analysis. The second stage, collection of the intensity data, is usually carried out by an automatic four-circle diffractometer. When a complete set of $Fo(hkl)$'s is given, the third stage begins. The number of atomic co-ordinates to be determined can be derived from Z and the space group. The positions of the heavy atoms are first deduced from Patterson maps. Positions of the lighter atoms are determined by successive Fourier syntheses of electron density and finally all the atomic co-ordinates so obtained are refined by least-squares methods.

The bond lengths and angles can be calculated on the basis of the final atomic co-ordinates. The errors in bond lengths and angles arise from errors in cell dimensions and atomic (fractional) co-ordinates. The former errors are much smaller than the latter. The latter arise from errors in intensity measurement. By statistical methods the standard deviations of bond lengths and angles can be estimated from the errors in intensity measurement. The standard deviations estimated in this way are usually tabulated with the bond lengths and angles in many published structural papers.

10 Neutron Diffraction

A beam of neutron of uniform velocity behaves like a wave of definite wavelength and can be diffracted by crystals. In a nuclear reactor, because of the principle of equipartition of energy, the neutrons have a Maxwellian distribution of energy. The thermal energy of a neutron at temperature T is $3kT/2$, where k is Bolzmann's constant. This must be equal to the kinetic energy, $mv^2/2$, where m is the mass of neutron.

Therefore,

$$mv^2/2 = 3kT/2,$$

and from de Broglie's relationship,

$$\lambda = h/mv,$$

where h is Planck's constant.

Thus

$$\lambda = h/\sqrt{3mkT} \tag{2.18}$$

The neutron beam from the reactor is received on a single crystal large enough to cover the entire beam. At any particular angle of incidence, θ, neutrons will be reflected if their wavelength satisfies the Bragg equation:

$$\lambda = 2d \sin \theta, \tag{2.19}$$

where d is the interplanar spacing. By selecting the angle, θ, appropriately, a beam of neutrons of the wavelength suitable for structural study can be obtained. With these monochromatic neutrons we can carry out the same kind of studies as with X-rays. A four-circle diffractometer is used for this purpose. The detecting counter is a proportional counter filled with ^{10}B-enriched boron trifluoride. For nearly all atoms it is atomic nuclei that are responsible for scattering neutrons. The interaction between atom and neutron is quite different in nature. Unlike X-ray scattering the scattering of neutrons by diamagnetic atoms is purely nuclear. Most atoms scatter neutrons equally well, within a factor of two or three. This is in contrast to the rapid increase with atomic number of the X-ray scattering amplitude. Thus neutron diffraction provides information on the location of the nuclei of light atoms like hydrogen to nearly the same accuracy as it does on other heavier atoms. In the study of co-ordination compounds, neutron diffraction is often carried out to ascertain the positions of light atoms in the presence of heavier ones.

Another important application is the study of bonding electrons in a molecule. Neutron diffraction gives the location of atomic nuclei, whereas X-ray diffraction provides electron density distribution in a molecule. Thus a combination of the two techniques will enable us to study the deformation of the electron cloud due to bonding more adequately than the X-ray diffraction study alone.

Although neutron diffraction gives us the results that are unobtainable with X-rays, it has some shortcomings. Firstly it is expensive: nuclear reactors are required. Secondly, small crystals cannot be used, since the diffracted beam is too weak to be measured. Usually a single-crystal specimen should have the dimension of several millimeters in each direction. Thirdly, strict monochromatisation may lead to a great loss of intensity. This will prevent very precise intensity measurement, resulting in the loss of accuracy in atomic co-ordinates.

11   An Example: X-Ray and Neutron Diffraction of $(-)_{589}$-tris(R-propylenediamine)cobalt(III) Bromide

As an illustration of the methods, the crystal structure determination of $(-)_{589}$-[Co(-pn)$_3$]Br$_3$ will be briefly sketched (Shintani, Sato and Saito, 1979). The compound was synthesized from Na$_3$[Co(CO$_3$)$_3$] and $(-)_{589}$-pn. The crystals were grown from an aqueous solution by slow evaporation. They are orange-red hexagonal needles. The crystal specimen was shaped into a sphere of radius 0.33 mm and used for X-ray intensity data collection.

The crystals are hexagonal[4] with unit cell dimensions: $a = b = 10.998(1)$ Å and $c = 8.567(1)$ Å at 25 °C.[5] The space group is $P6_3$.[6] The observed density of the crystal was 1.91 g cm^{-3}. By means of Eq. (2.17) the number of formula units in the unit cell was determined to be 2. The intensity data were collected on an automatic four-circle diffractometer. In total, 1,423 distinct reflexions were collected and used for the solution of the structure and refinement of atomic co-ordinates. The positions of the heavy atoms, Co and Br in this case, could be deduced from the Patterson synthesis. Positions of the lighter atoms were determined by successive Fourier syntheses of electron density. The atomic co-ordinates thus obtained were refined by the least-squares method. The final R value became 0.056 for the 1423 ob-

---

4   The angle between the a and b axes is 120°. The c axis is perpendicular to the plane formed by the a and b axes.
5   Standard deviations to the last digit are usually given in parentheses.
6   For notation of the space group readers may consult International Tables for X-ray Crystallography, Vol. I (1959).

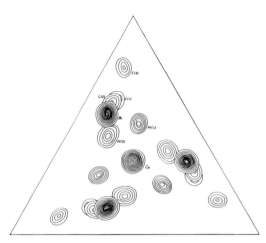

Fig. 2.8. Composite three dimensional Fourier diagram of electron density in $(-)_{589}$-$[Co(R\text{-}pn)_3]Br_3$. Contours around the cobalt and bromine atoms are drawn at intervals of $10\ e\ Å^{-3}$, and those for other lighter atoms at $2\ e\ Å^{-3}$

served reflexions. Fig. 2.8 shows the final electron density map of $(-)_{589}$-$[Co(R\text{-}pn)_3]$ Br_3. This diagram is composed of sections through each atomic centre. All the atoms except hydrogen come out spherically in the diagram. Peak heights increase with increasing atomic number. Peaks due to hydrogen atoms are not drawn, since the peak height is only $1\ e\ Å^{-3}$. The corresponding packing diagram is shown in Fig. 2.10.

The crystal specimen used for the collection of the neutron intensity data was shaped into a sphere of 3.0 mm in radius. The relative intensities of a total of 364 independent reflexions were measured. The structure was refined by least-squares methods. The final R value was 0.050 for the 364 observed reflexions. Table 2.1 compares the scattering amplitudes of the atoms in $[Co(R\text{-}pn)_3]Br_3$ for X-rays and neutrons. A composite three dimensional Fourier synthesis corresponding to that shown in Fig. 2.8 is presented in Fig. 2.9. Owing to the difference in scattering amplitudes for X-rays and neutrons the peaks due to nitrogen and carbon atoms are even higher than the cobalt and bromine peaks. Hydrogen peaks come out clearly, but they are indicated by broken lines to show their negative scattering amplitudes. An interesting feature is that the hydrogen atoms attached to the methyl carbon are dis-

Table 2.1. Scattering amplitudes for X-rays and neutrons

Element	Atomic number	X-ray scattering amplitudes in electrons		Neutron scattering amplitudes in 10^{-12} cm
		$\sin\theta/\lambda = 0$	$\sin\theta/\lambda = 0.50^a$	
H	1	1.0	0.07	−0.372
C	6	6.0	1.7	0.665
N	7	7.0	1.9	0.94
Co	27	27.0	12.2	0.25
Br	35	35.0	18.3	0.67

a   in $Å^{-1}$

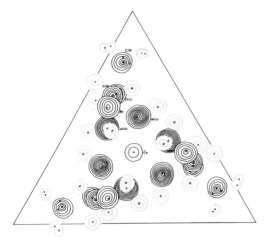

Fig. 2.9. Composite three dimensional Fourier diagram of neutron scattering density in [Co(R-pn)$_3$Br$_3$. Contours are drawn at intervals of 5 × 10^{12} cm Å^{-3}

ordered. There are six negative peaks around each methyl carbon atom with a half peak height of other hydrogen peaks. Such arrangement of peaks seem to indicate that the methyl group takes two alternative sets of positions with equal probability. Neutron diffraction does not tell us whether this is a dynamical pseudorotation of the methyl group or simply the statistical orientational disorders. A packing diagram of $(-)_{589}$-[Co(R-pn)$_3$]Br$_3$ is shown in Fig. 2.10.

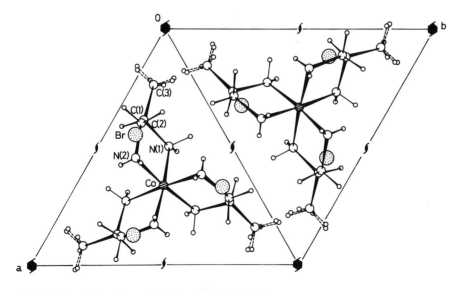

Fig. 2.10. A packing diagram of $(-)_{589}$[Co(R-pn)$_3$]Br$_3$

12 Anomalous Scattering

X-ray methods can give the detailed geometry of a complex, interatomic distances and bond angles; however, it is not possible by the normal X-ray method to decide whether the optically active complex has a particular configuration or its mirror image. This is because diffracted intensities are usually measured under such conditions that the X-ray wavelength used is nowhere near the absorption edge of any atom in the crystal. A typical absorption spectrum of an atom for X-rays is shown in Fig. 2.11. The absorption coefficient increases gradually with increase in λ and it decreases abruptly and begins to increase again. These sudden discontinuities in atomic absorption coefficient are known as absoprtion edges. They can be understood in terms of the electronic structure of the atom. Within the atom, electrons exist in definite energy states and the absorption edges correspond to energies hc/λ which are just sufficient to eject an atomic electron from the atom. Thus absorption edges are classified according to the electrons ejected by the incident X-rays: K, L, M etc. When the X-ray wavelength lies near one of the absorption edges of any atom in the crystal, X-rays are scattered anomalously from the particular atom. Let us first consider what happens when the incident X-rays are scattered anomalously.

The physical basis of the anomalous scattering can be seen by carrying out a simple experiment. What we wish to do is to observe how one oscillator can affect others of different frequencies. Figure 2.12 illustrates the experiment. The black circle represents a heavy plumb suspended by a horizontal string, tightly supported at both ends. To the same string are attached other pendulums of different lengths, with less heavy plumbs. One pendulum should have the same length as the master pendulum. If one makes the master pendulum oscillate in a plane perpendicular to the horizontal string, it will oscillate slowly and then the other pendulums will also start to oscillate, because the pendulums are coupled together. One will find that the pendulum with the same length (hence the same natural frequency) as the master pendulum will have a far greater amplitude than any of the others. This is the well known phenomenon of resonance. If one looks at the other pendulums more closely, one may notice that when all their motions become stationary, they can be seen to be vibrating with the same frequency as the master pendulum. But those that are shorter than the master pendulum are vibrating out of phase. If the master pendulum

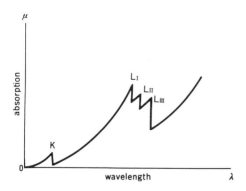

Fig. 2.11. Variation of linear absorption coefficient with wavelength for a typical element

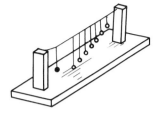

Fig. 2.12. Eight pendulums are shown. The black one is heavy: the others are lighter

is considered to be the incident X-ray beam, and the driven oscillators to be the atomic electrons in the various energy levels of the scattering atoms, what happens in the pendulum system exactly represents what happens in the atoms. The scattering by all kinds of electrons is in phase, since these electrons oscillate with the same frequency as the incident beam. In the region near to resonance odd changes in phase can take place: all the electrons do not now scatter in phase and the scattered X-rays have different phase angles. The anomalous scattering will be discussed in the next section on the basis of a classical model.

13 The Forced, Damped Oscillations of an Electron

In the classical theory, the scattered radiation of unchanged wavelength is produced by the forced oscillations of the electrons. If we consider one atomic electron, moving independently of all others, then its oscillational motion in the alternating electric field of an incident electromagnetic wave, $E_0 \exp(i\omega t)$, can be described by an equation of the form

$$\frac{dx^2}{dt^2} + \gamma \frac{dx}{dt} + \omega_s^2 x = - \frac{eE_0}{m} \exp(i\omega t) \tag{2.20}$$

In Eq. (2.20), m is the electronic mass, $-e$ the electronic charge and ω_s is the resonance frequency of an electron weakly bound by the nucleus, thus the restoring force being given by $-m\omega_s^2 x$. γ in the second term of the left-hand side of Eq. (2.20) is a positive constant and $-m\gamma(dx/dt)$ stands for the loss of electronic energy by resonance absorption which is represented as a damping force proportional to the velocity of the electron. From (2.20), the dipole moment of this oscillating electron P is given by

$$P = -ex = \frac{e^2}{m} \frac{E_0 \exp(i\omega t)}{\omega_s^2 - \omega^2 + i\gamma} \tag{2.21}$$

Classical electromagnetic theory tells us that the oscillating electric dipole emits an electromagnetic wave of the same frequency. The value of the electric vector of the scattered wave observed on a plane perpendicular to the oscillating dipole at a distance r (large compared to the wavelength) is given by $P(t - r/c) \times \omega^2/cr^2$, where $P(t - r/c)$

is the electric moment at time $t - r/c$ and c is the light velocity in vacuum. Thus the amplitude of oscillation at unit distance A can be written as

$$A = \frac{e^2}{mc^2} \times \frac{\omega^2 E_0}{\omega_s^2 - \omega^2 + i\gamma} \qquad (2.22)$$

For a free electron (no binding force by the nucleus and no resonance absorption) one would have $\omega_s = 0$ and $\gamma = 0$. Its scattering amplitude A' becomes

$$A' = - (e^2/mc^2)E_0 \qquad (2.23)$$

The negative amplitude indicates an oscillation of the electron with the same frequency as the incoming wave but π out of phase. By writing (2.23) as

$$A' = (e^2/mc^2) \exp i\pi$$

the phase shift of π with respect to the incoming wave $E = E_0 \exp(i\omega t)$ is made more evident.

The atomic scattering factor ϕ of the oscillating dipole is the scattering amplitude measured in terms of that of a free electron at the same experimental condition.

$$\phi = A/A' = \frac{\omega^2}{\omega^2 - \omega_s^2 - i\gamma} \qquad (2.24)$$

If we separate ϕ into the real and imaginary components and put

$$\phi = \phi' + i\phi'' \qquad (2.25)$$

we obtain

$$\phi' = \frac{\omega^2(\omega^2 - \omega_s^2)}{(\omega^2 - \omega_s^2)^2 + \gamma^2\omega^2} \qquad (2.26)$$

$$\phi'' = \frac{\gamma\omega^3}{(\omega^2 - \omega_s^2)^2 + \gamma^2\omega^2} \qquad (2.27)$$

The most tightly bound electrons which therefore have the largest ω_s value are the K electrons. However, even for these ω is much greater than ω_s and their behaviour is not very much different from that of a free electron. Hence ϕ'' is a small quantity. We have so far considered only one electron. In an actual atom, however, there are many atomic electrons in the various energy levels. Thus the real atomic scattering factor is given by a sum of terms like that given by Eq. (2.25) for all atomic electrons, namely,

$$f = \sum_j \phi'_j + i \sum_j \phi''_j = f_0 + \Delta f' + i \Delta f'' \qquad (2.28)$$

where f_0 is the scattering factor at infinite wavelength, $\Delta f'$ is the small correction for the real part of the scattering factor at λ and $\Delta f''$ is the imaginary part of it. f can be further rewritten as:

$$f = \sqrt{(f_0 + \Delta f')^2} + \Delta f'' \exp(i\beta) \tag{2.29}$$

$$\tan \beta = \frac{\Delta f''}{(f_0 + \Delta f')} \tag{2.30}$$

where β is the phase shift on anomalous scattering. From Eq. (2.27) we find $\Delta f''$ is a positive quantity, hence β is a small positive angle. With the time factor $\exp(i\omega t)$ this means that the anomalous phase shift on scattering has the effect of advancing the wave. It should not be assumed that this rather crude theory based on classical ideas gives good agreement with experiment but it does give a reasonable qualitative picture. The theory of anomalous scattering was first developed by Hönl (1933, a and b). His first paper was a classical treatment and the second one was a quantum mechanical treatment. More advanced quantum field theory gives essentially the same result (see for example, Sakurai, 1967). The way in which $\Delta f'$ and $\Delta f''$ depend on the incident wavelength in relation to the absorption edge is illustrated in Fig. 2.13. The values of $\Delta f'$ and $\Delta f''$ were calculated by quantum mechanical theory. $\Delta f''$ is always positive when the wavelength of incident radiation is shorter than the absorption edge wavelength, whereas it is zero when it is longer than the absorption edge.[7]

Calculated values of $\Delta f'$ and $\Delta f''$ are tabulated in Vols. III and IV of International Tables for X-ray Crystallography (1962, 1974). Experimental verification of these values is rather fragmentary, but the agreement is generally good, where experimental data exist.

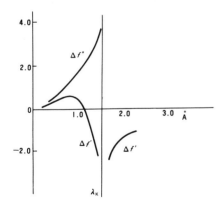

Fig. 2.13. Variation of $\Delta f'$ and $\Delta f''$ of cobalt with wavelength near K absorption edge

---

7 Classical theory predicts a finite value of $\Delta f''$ when $\lambda > \lambda_K$ [see Eq. (2.27)].

14 Direct Consequences of Anomalous Scattering

Let us consider a simple structure consisting of atom A and atom B. These atoms lie in periodically alternating planes. If the two planes A and B are arranged with equal spacing as shown in Fig. 2.14 (a), we cannot distinguish the upward and downward directions. However, if they are arranged with non-equal spacing between adjacent layers [Fig. 2.14 (b)], we can distinguish the two directions by the sequence of A and B and the distance between them:

	distance from A to B	distance from B to A
upward	long	short
downward	short	long

─────────────── A

─────────────── A
─────────────── B

─────────────── B

─────────────── A (a)

─────────────── A
─────────────── B (b)

Fig. 2.14. Polar and non-polar direction in the structure **(a)** non-polar and **(b)** polar

Such a structure is called polar as against the non-polar structure shown in Fig. 2.14 (a).

The polar and non-polar directions coexist in a polar structure. Figure 2.15 shows the atomic arrangement in crystals of zincblende (sphalerite), a modification of zinc sulphide. The zinc atom is surrounded by four sulphur atoms at the corners of a tetrahedron, the sulphur being similarly surrounded by four metal atoms. The arrangement of the atomic planes parallel to (100) and (111) are shown in Fig. 2.15 (b) and (c). One can easily see that the former arrangement is non-polar, while the latter is polar, the Zn and S planes occuring alternately in pairs. The polar nature of the atomic arrangement is reflected in its crystal habit. Figure 2.16 illustrates an

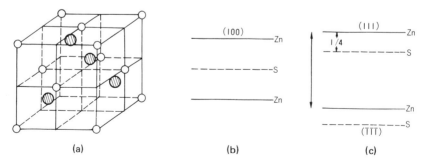

(a)   (b)   (c)

Fig. 2.15. The structure of zincblende (ZnS), showing the atomic arrangement **(a)**, atomic arrangement in the planes parallel to (100) **(b)** and to (111) **(c)**

Fig. 2.16. Zincblende

ideally well-formed single crystal of zincblende. It is a modified tetrahedron, showing additional facets on each corner. A small facet and a large face on the opposite side of the crystal are exactly parallel, corresponding to an atomic plane and its rear: (111) and $(\overline{1}\overline{1}\overline{1})$ or their equivalents. One face is usually shinier than the other; the former becomes positively charged when the crystal is compressed to the direction perpendicular to the face. The etch figures and velocity of crystal growth are different, too. Let us consider the reflexions from (111) and $(\overline{1}\overline{1}\overline{1})$. For the normal diffraction the intensities of 111 and $\overline{1}\overline{1}\overline{1}$ are equal and hence indistinguishable. The situation is illustrated in Fig. 2.17. The wave scattered by the plane of S atoms is delayed by $360° \times (1/4)$ $(=90°)$ with respect to that scattered by the plane of Zn atoms for the 111 reflexion, while for the counter reflexion $\overline{1}\overline{1}\overline{1}$, the phase change is $270°$ $(= -90°)$. The figure shows how the Zn and S contributions combine to give the total wave amplitude $|F(111)|$ and $|F(\overline{1}\overline{1}\overline{1})|$, respectively. We have

$$|F(111)| = |F(\overline{1}\overline{1}\overline{1})|$$

hence

$$I(111) = I(\overline{1}\overline{1}\overline{1})$$

The intensities of reflexions from 111 and $\overline{1}\overline{1}\overline{1}$ are equal and indistinguishable. Generally $|F(hkl)| = |F(\overline{hkl})|$ for normal scattering, as will be shown in the next section. This is called Friedel's rule.

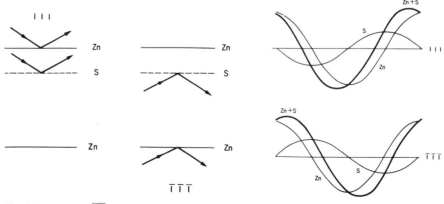

Fig. 2.17. 111 and $\overline{1}\overline{1}\overline{1}$ reflections from zincblende

The K absorption edge wavelength of zinc lies at 1.281 Å. Thus the incident X-rays are expected to be scattered anomalously by the zinc atom, if the wavelength is selected to be close to (but necessarily shorter than) 1.281 Å. $WL\beta_1$ (λ = 1.2792 Å) or Au $L\alpha_1$ (λ = 1.2738 Å) is suitable for this purpose. This classical experiment on anomalous scattering was carried out by Nishikawa and Matsukawa (1928) and by Coster, Knol and Prins (1930). The anomalous phase shift on scattering has the effect of advancing the wave from Zn relative to the wave from S according to Eq. (2.28). It was calculated that for Au $L\alpha_1$ radiation, the relative phase on scattering by zinc would be advanced by 10.5°. This small phase shift results in a stronger reflexion of Au $L\alpha_1$ radiation from the $(\overline{1}\overline{1}\overline{1})$ face than the (111) as illustrated in Fig. 2.18. In this way the polar direction can be distinguished by means of X-ray anomalous scattering. The shinier face was found to correspond to (111) in Fig. 2.16. Knowledge of crystallographic polarity is important, for instance, in the theory of piezoelectricity and also in solid-state electronics, since this polarity influences the band bending near the surface and thus such properties as photoemission depend upon the polarity (James, Antypas, Edgecombe, Moon and Bell, 1971). Brongersma and Mul (1973) verified the above assignment of the polar direction by a different method. They studied the opposite faces of ZnS by measuring the energy of noble gas ions scattered from the surface. Noble gas ions, like Ne^+, which are back-scattered from a surface lose an amount of energy characteristic for the mass of the surface atom with which they collide. The two surfaces of ZnS were analysed by specular reflection of Ne^+ ions with an incident energy of 1,000 eV in ultra high vacuum. Energy of the scattered Ne^+ ions was determined by mass spectrometry. The result showed that the $(\overline{1}\overline{1}\overline{1})$ face is practically covered by sulphur atoms, while no sulphur was detected on the opposite (111) face, in agreement with the assignment based on anomalous scattering.

We can extend this result to more complicated structures, comprising many atoms. In addition to A and B, there are more atoms, C, D, E, etc. As mentioned in section 5, the total wave amplitude from hkl is a sum of the waves scattered from each atomic plane with appropriate phase shift depending on the relative distance from the reference atomic plane. We still find that the total amplitude of the wave diffracted from hkl is the same as that diffracted from $\overline{hkl}$ for normal diffraction

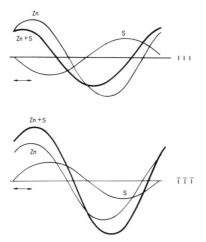

Fig. 2.18. 111 and $\overline{1}\overline{1}\overline{1}$ reflections from ZnS when X-rays are scattered anomalously from the zinc atom. Anomalous phase shift is exaggerated in the figure and is shown by an arrow

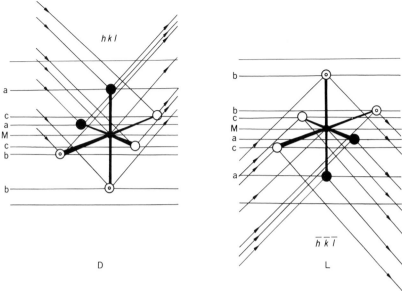

Fig. 2.19. Equivalence of $F_D\ (hkl)$ and $F_L\ (\overline{hkl})$

(f_j: real) and the phase angles are equal in magnitude and opposite in sign, no matter whether the structure is polar or not. Now let us consider a dissymmetric structure D and its enantiomeric structure L. In Fig. 2.19 the two structures are represented by the enantiomeric octahedral complex [$Ma_2b_2c_2$]. It can easily be seen that the reflexion $\overline{hkl}$ from L is equivalent to hkl reflexion from D: the total amplitudes as well as the phase shifts are the same[8].

Accordingly we have:

$$F_D(hkl) = F_L(\overline{hkl}) \tag{2.31}$$

In the same way we find

$$F_L(hkl) = F_D(\overline{hkl}) \tag{2.32}$$

By Friedel's rule

$$|F_D(hkl)|^2 = |F_D(\overline{hkl})|^2 = |F_L(hkl)|^2 = |F_L(\overline{hkl})|^2$$

This result indicates that what we obtain from the D and L crystals by normal X-ray diffraction is two sets of intensity data that are equal to each other, thus we cannot determine the absolute configuration. When anomalous scattering occurs in one of the atoms in the crystal, Friedel's rule is violated and if, for instance,

---

8   The hkl and $\overline{hkl}$ can be assigned unambiguously for D and L crystals, as far as the same set of co-ordinate systems, say, a right-handed one, is used throughout.

$$|F_D(hkl)|^2 > |F_D(\overline{hkl})|^2,$$

then

$$|F_L(hkl)|^2 < |F_L(\overline{hkl})|^2$$

by means of Eqs. (2.31) and (2.32). In other words, the inequality relations observed for one structure are the reverse of those found for its enantiomorph. This is the basis of the determination of absolute configuration by the anomalous scattering technique. The intensities can be calculated by assuming a particular enantiomeric configuration for the complex and the result can be compared with the observation. If the intensity relations are the reverse of those observed, then the inverted configuration represents the correct absolute configuration.

15 Complete Expressions for Friedel's Rule and its Breakdown by Anomalous Scattering

Let us formulate what we have discussed in the previous section. For normal diffraction we have

$$F(hkl) = \sum_j f_j \exp\left[2\pi i(hx_j + ky_j + lz_j)\right]$$

where f_j's are all real quantities. For $\overline{hkl}$ reflexion

$$F(\overline{hkl}) = \sum_j f_j \exp\left[-2\pi i(hx_j + ky_j + lz_j)\right]$$

Therefore $|F(hkl)| = |F(\overline{hkl})|$ or $I(hkl) = I(\overline{hkl})$. This is Friedel's rule.

Consider a unit cell containing one dissymmetric molecule: an octahedral complex, $[Ma_2b_2c_2]$ (Fig. 2.20). The structure factors for the hkl and $\overline{hkl}$ reflexions can be written

$$F_D(hkl) = \sum_j f_j \exp\left[2\pi i(hx_j + ky_j + lz_j)\right] = A(hkl) + i\,B(hkl) \tag{2.33}$$

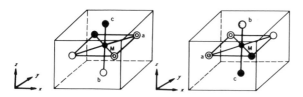

Fig. 2.20. Dissymmetric molecule in a unit cell (a) and the structure in (a) inverted (b)

$$F_D(\overline{hkl}) = \sum_j f_j \exp[-2\pi i(hx_j + ky_j + lz_j)] = A(hkl) - i\,B(hkl) \qquad (2.34)$$

where

$$A(hkl) = \sum_j f_j \cos 2\pi(hx_j + ky_j + lz_j)$$

and

$$B(hkl) = \sum_j f_j \sin 2\pi(hx_j + ky_j + lz_j)$$

If the structure shown in Fig. 2.20 (a) is inverted, the enantiomorphous structure illustrated in Fig. 2.20 (b) is obtained. The atomic co-ordinates in this inverted structure are $(-x_j, -y_j, -z_j)$, where x_j, y_j and z_j are those of the structure shown in Fig. 2.20 (a). The structure factor for the inverted structure can be written:

$$F_L(hkl) = \sum_j f_j \exp[-2\pi i(hx_j + ky_j + lz_j)]$$

$$F_L(\overline{hkl}) = \sum_j f_j \exp[2\pi i(hx_j + ky_j + lz_j)]$$

Accordingly we have

$$F_D(hkl) = F_L(\overline{hkl})$$

and

$$F_D(\overline{hkl}) = F_L(hkl),$$

the same result as that derived by inspection of Fig. 2.19. This is called Bijvoet's relation (Bijvoet, Peerdeman and van Bommel, 1951). If the wavelength of the incident X-rays is appropriately chosen in such a way that it is scattered anomalously by one kind of atom, say M, in the unit cell, the scattering factor for this atom is represented by a complex quantity (Eq. 2.28):

$$f_M = f_{0M} + \Delta f'_M + i\Delta f''_M \qquad (2.35)$$

For simplicity, we assume that there is only one M atom in the unit cell.
By inserting Eq. (2.35) into (2.33),

$$F_D(hkl) = \sum_j f_j \exp[2\pi i(hx_j + ky_j + lz_j)]$$

$$= (f_{0M} + \Delta f'_M + i\Delta f''_M)\exp[2\pi i(hx_M + ky_M + lz_M)]$$

$$+ \sum_j{}' f_j \exp[2\pi i(hx_j + ky_j + lz_j)] \qquad (2.36)$$

where Σ' means that the summations are taken over all the atoms except M in the unit cell.

Eq. (2.36) can be written

$$F_D(hkl) = A(hkl) + i\,B(hkl)$$

$$A(hkl) = (f_{0M} + \Delta f'_M)\cos\left[2\pi(hx_M + ky_M + lz_M)\right] - \Delta f''_M \sin\left[2\pi(hx_M + ky_M + lz_M)\right]$$
$$+ \Sigma'_j f_j \cos\left[2\pi(hx_j + ky_j + lz_j)\right]$$

$$B(hkl) = (f_{0M} + \Delta f'_M)\sin\left[2\pi(hx_M + ky_M + lz_M)\right] + \Delta f''_M \cos\left[2\pi(hx_M + ky_M + lz_M)\right]$$
$$+ \Sigma'_j f_j \sin\left[2\pi(hx_j + ky_j + lz_j)\right]$$

Accordingly we have

$$|F_D(hkl)| \neq |F_D(\overline{hkl})| \tag{2.37}$$

The situation is illustrated on an Argand diagram in Fig. 2.21. Note that the vector $\overline{\Delta f''_M}$ is perpendicular to $\overline{f_{0M} + \Delta f'_M}$, since the argument of the former is greater by $90°$ than the latter (advance in phase).

First application of anomalous scattering for the determination of the absolute stereochemical configuration of an organic molecule was made by Bijvoet, Peerdeman and van Bommel in 1951. They studied sodium rubidium d-tartrate tetrahydrate, $NaRbC_4H_4O_6 \cdot 4H_2O$ using zirconium Kα radiation. The K absorption edge wavelength of rubidium atom was 0.814 Å and the wavelength of Zr Kα radiation was 0.784 Å. Accordingly the radiation was scattered anomalously from the rubidium atom and the difference in intensities of a reflexion and its counter reflexion was clearly discernible on the diffraction photographs. The absolute configuration of d-

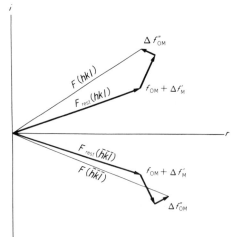

Fig. 2.21. The breakdown of Friedel's rule by an anomalous scatterer. F_{rest} represents the contribution from the remaining atoms

Table 2.2. Observed and calculated intensities

Plane	Calculated		Observed	
hkl	$I(hkl)$	$I(hk\bar{l})$	$(+)_{589}$-isomer	$(-)_{589}$-isomer
112	1	133	<	>
212	3,700	3,450	>	<
312	323	440	<	>
412	18	31	<	>
612	42	39	>	<
122	71	150	<	>
222	38	157	<	>
322	142	87	>	<

tartaric acid was determined as R,R. The result agreed, as it happened, with Emil Fischer's convention which had been arbitrarily chosen.

Three years later, Saito, Nakatsu, Shiro and Kuroya determined the absolute configuration of tris(ethylenediamine)cobalt(III) ion by the same technique and they first demonstrated the reversal of inequality for a pair of enantiomorphous structures by employing crystals of $(+)_{589}$- and $(-)_{589}$-$[Co(en)_3]_2$ $Cl_6 \cdot NaCl \cdot 6H_2O$ and such inequality relations were unequivocally shown to be due to the effect of anomalous scattering and not to some other effect such as difference in surface roughness or absorption. The K absorption edge of the cobalt atom lies at 1.608 Å. In the experiment the crystal setting was adjusted by using Fe Kα radiation ($\lambda = 1.937$ Å; hence normal scattering) in such a way that the reflexions from hkl and $hk\bar{l}$ (the latter being equivalent to $\overline{hkl}$ by the symmetry requirement of the space group) showed equal intensity. Then Cu Kα radiation ($\lambda = 1.542$ Å), which was scattered anomalously by the cobalt atoms was used to take diffraction photographs: $\Delta f'$ -2.5, $\Delta f''$ 3.6 so that the phase advance (β) is $8.3°$. Table 2.2 shows the calculated intensities and observed relations for the two enantiomeric crystals. As seen from the table the inequality relations are reversed for $(+)_{589}$- and $(-)_{589}$-crystals. The calculated intensities based on the set of co-ordinates representing Λ-$[Co(en)_3]^{3+}$ agree with the observed intensity relations for $(+)_{589}$-isomer. Thus the dextrorotatory isomer was found to have Λ absolute configuration.

Figure 2.22 (a) and (b) show the effect of anomalous scattering. These are Weissenberg photographs of $(-)_{589}$-$[Co(S,S\text{-chxn})_3]Cl_3 \cdot 5H_2O$. Figure 2.22 (a) was taken with Fe Kα radiation. Accordingly no effect of anomalous scattering is observed. Intensity distributions of a number of diffraction spots on a characteristic U shaped row-line are symmetric with respect to its centre. A pair of reflexions symmetrically located on a row-line corresponds to a reflexion hkl and its counter reflexion $\overline{hkl}$ (strictly speaking its equivalent). Therefore, we recognize that the Friedel's rule is obeyed. On the other hand, (b) was taken by Cu Kα radiation. X-rays are then scattered anomalously by the cobalt atom. Copious fluorescent radiation darkens the background of the film. The corresponding reflexions on a row-line are no longer equal in intensity and Friedel's rule is violated. If a crystal of the Δ-isomer is used under

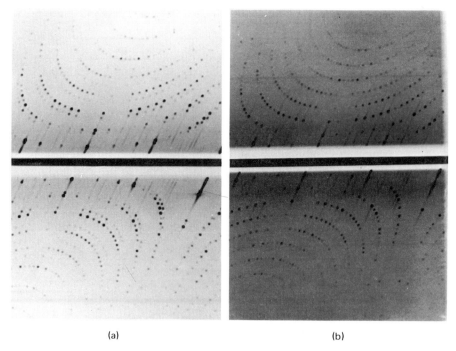

(a)   (b)

Fig. 2.22. Weissenberg photographs showing the effect of anomalous scattering. $(-)_{589}\Lambda$-
[Co(S,S-chxn)$_3$]Cl$_3 \cdot$ 5H$_2$O, a-axis rotation, 0-th layer. **(a)** Taken with Fe Kα radiation
(λ = 1,937 Å). **(b)** Taken with Cu Kα radiation (λ = 1,542 Å) (Saito, 1974)

the same experimental conditions, the observed intensity will be inverted. Recent accurate measurement of the intensities of Bijvoet pairs show quantitative agreement with
the calculated values. Table 2.3 compares the observed and calculated intensities of
Bijvoet pairs for which the observed $|F(hkl)|$ and $|F(\overline{hkl})|$ of $(+)_{589}$-[Co(linpen)]
[Co(CN)$_6$] · 3H$_2$O differed by more than 15%. The agreement in the table indicates
that the complex ion $(+)_{589}$[Co(linpen)]$^{3+}$ has the absolute configuration $\Lambda\Lambda\Lambda\Delta$
(Sato and Saito, 1975).

Table 2.3. Observed and calculated structure amplitudes of Bijvoet pairs: $(+)_{589}$[Co(linpen)]
[Co(CN)$_6$] · 3H$_2$O

hkl	$|Fo|$	$|Fc|$	hkl	$|Fo|$	$|Fc|$
610	33.2	31.8	421	43.7	40.5
$\overline{6}$10	27.0	26.0	$\overline{4}2\overline{1}$	36.7	33.3
630	34.0	32.2	621	41.6	39.6
$\overline{6}$30	27.2	24.6	$\overline{6}2\overline{1}$	33.6	32.3
411	11.1	14.0	821	16.7	18.7
$\overline{4}1\overline{1}$	35.0	35.8	$\overline{8}2\overline{1}$	28.0	28.1
811	32.3	31.3	031	15.0	7.6
$\overline{8}1\overline{1}$	22.3	24.2	$0\overline{3}\overline{1}$	21.8	24.2

Okaya, Saito and Pepinsky (1955) showed that a synthesis analogous to Patterson function:

$$P_s(UVW) = 1/V \ \Sigma\Sigma\Sigma \ |F(hkl)|^2 \ \sin\left[2\pi(hU + kV + lW)\right] \qquad (2.38)$$

gives, when summed up for structure amplitudes of dissymmetric crystals measured under anomalous scattering conditions, the distribution of interatomic vectors between anomalous scatterers and normal scatterers, including the absolute sence of the vectors; positive peaks represent vectors from anomalous scatterers to normal scatterers, negative peaks represent vectors from normal to anomalous scatterers. This synthesis can be used to determine the absolute configuration of the molecule. Iwasaki (1974) showed that certain dissymmetric structures containing two or more kinds of anomalous scatterers cannot violate Friedel's rule and their absolute configuration can never be established by the usual anomalous scattering technique. No structure of this type, however, seems to have been recorded so far, though one might be found in the future.

16 The Use of Internal Reference Centre

A known absolute configuration in one asymmetric centre can be utilized to deduce that of other centres in the structure. In this case the whole structure is determined in such a way that the atomic group of known absolute configuration gives the correct spacial arrangement. The method was first suggested by Mathieson (1956) and is now widely used, since the number of moieties of known absolute configuration has increased considerably. For instance, the absolute configuration of $(-)_{589}[Co(NO_2)_2$ $(ox)(NH_3)_2]^-$ was determined with reference to the known absolute configuration of $\Delta\text{-}(-)_{589}[Co(NO_2)_2(en)_2]^+$ by solving the crystal structure of $(-)_{589}\text{-}[Co(NO_2)_2$ $(en)_2](-)_{589}[Co(NO_2)_2(ox)(NH_3)_2]$ (Shintani, Sato and Saito, 1976). In some complexes the moiety of known absolute configuration is an intrinsic part of the complex for which the absolute configuration is to be determined. Such was the case with the $(+)_{589}\text{-}[Co(R\text{-}pn)_3]^{3+}$ ion (Kuroda and Saito 1974). The absolute configuration was determined to be $\Lambda\text{-}(ob_3)\text{-}fac$-isomer with reference to R-pn by the structure determination of $(+)_{589}[Co(R\text{-}pn)_3][Co(CN)_6] \cdot 2H_2O$ (see p. 60).

17 Neutron Anomalous Scattering

Some of the nucleides like ^{113}Cd, ^{149}Sm, ^{157}Gd, ^{153}Eu, etc., scatter neutrons anomalously in the thermal energy region, where the neutron wavelength is comparable to that of X-rays used for structure analysis. Thus the absolute configuration can be determined in the same way as the X-ray anomalous scattering technique. The phenomenon of anomalous scattering, i.e. resonance scattering is so marked in the neutron case in comparison with X-rays that the dispersion terms are generally one order of magnitude greater than the normal scattering amplitudes. Furthermore, the behaviour of the real and imaginary components of the scattering amplitudes with the wavelength is quite different from that in X-rays. The physical process leading to resonance behaviour is through the formation of a compound nucleus:

$$n + {}^{113}Cd \rightarrow {}^{114}Cd \quad\begin{cases} \nearrow & n + {}^{113}Cd \quad \text{coherent} \\ \searrow & \gamma + {}^{114}Cd \quad \text{incoherent} \end{cases}$$

Both processes are possible and the first gives coherent neutrons and the second incoherent. For the first process the scattering length for neutrons is given by

$$b_0 = b' + i\,b'' \tag{2.39}$$

The absolute configuration of aqua(S-glutamato)cadmium(II) hydrate was determined by neutron diffraction (Flook, Freeman and Scudder, 1977). The nuclide ^{113}Cd, which is present in naturally occuring Cd to the extent of 12.3%, was used as an anomalous scatterer for neutron wavelength of about 0.7 Å. In this case, the neutron scattering length is

$$(0.38 + i\,0.12) \times 10^{-12} \text{ cm}$$

(Peterson and Smith, 1962).

Appendix II-1

In Fig. 2.23 OX, OY and OZ are crystal axes. Consider a set of planes hkl; one passes through the origin, O, the next intercepts with the axes at A, B and C. Then

$$OA = a/h, \quad OB = b/k \quad \text{and} \quad OC = c/l.$$

Drop a normal from O to the plane ABC. The length ON ($= d(hkl)$) is the interplanar spacing. Let $A'B'C'$ be a plane parallel to ABC and through a point with fractional co-ordinates (x_2, y_2, z_2). The length ON' is $d'(hkl)$ in Fig. 2.5. The plane $A'B'C'$ makes intercepts of (a/h) $(d'(hkl)/d(hkl))$, (b/k) $(d'(hkl)/d(hkl))$ and (c/l) $(d'(hkl)/d(hkl))$ on the axes.

The equation of this plane can be written as:

$$\frac{ax}{\dfrac{a}{h}\dfrac{d'(hkl)}{d(hkl)}} + \frac{by}{\dfrac{b}{k}\dfrac{d'(hkl)}{d(hkl)}} + \frac{cz}{\dfrac{c}{l}\dfrac{d'(hkl)}{d(hkl)}} = 1$$

The point (x_2, y_2, z_2) is on this plane. Accordingly we have

$$\frac{ax_2}{\dfrac{a}{h}\dfrac{d'(hkl)}{d(hkl)}} + \frac{by_2}{\dfrac{b}{k}\dfrac{d'(hkl)}{d(hkl)}} + \frac{cz_2}{\dfrac{c}{l}\dfrac{d'(hkl)}{d(hkl)}} = 1 \qquad \therefore \quad hx_2 + ky_2 + lz_2 = d'(hkl)/d(hkl)$$

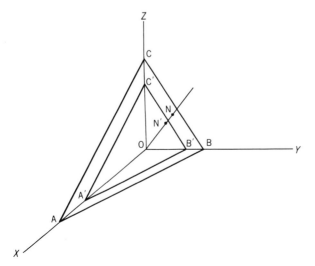

Fig. 2.23. See text

Chapter III Conformational Analysis

1 Introduction

Conformational analysis is usually concerned with the analysis of the physical and chemical properties of a molecule in terms of its conformational structure. It enables us to understand more deeply the properties of metal chelate complexes and the interactions of metal ions with ligands. We can predict the structure of unknown chelate complexes, estimate the strain energies of a series of conformers and predict the formation ratios with reasonable certainty and finally afford an explanation for the stereoselectivity from a detailed knowledge of the conformational features.

The concept of conformational analysis is well established in the field of organic chemistry, but its application to co-ordination compounds lags somewhat behind. Mathieu first attempted to explain why one diastereoisomer of cis-$[CoX_2(pn)_2]$ was formed preferentially (1944). He tried to calculate the energy difference between the diastereoisomers by considering London dispersion forces. Fifteen years later, Corey and Bailar published their classical paper (1959). These authors discussed the stabilities of $trans$-$[CoCl_2(pn)_2]^+$ and $[Co(en)_3]^{3+}$ isomers and showed the equatorial preference of substituted methyl groups in Co(pn) rings on the basis of calculations of non-bonded H...H interactions. In the 1960's a number of more advanced papers on conformational analysis of co-ordination compounds appeared. They are summarised in excellent review articles (Hawkins 1971; Buckingham and Sargeson 1971; Niketić and Rasmussen, 1977).

It is not intended in this chapter to give a comprehensive review of conformational analysis but to describe an outline of the method, hoping that it might assist the reader to comprehend the discussions on each particular complex in Chapter IV.

In this approach a molecule is considered as a system of point atoms held together in a particular spacial arrangement. When interatomic forces are balanced the molecular conformation is said to be in an equilibrium. The potential function of the system is written down in terms of various parameters defining the molecular geometry. The function is then minimized to obtain equilibrium conformation. The potential energy functions now used are largely empirical or semi-empirical containing several parameters which are adjusted to give the best fit to the observable molecular properties. In fact, the present state of theoretical knowledge is such that empirical functions often give a more realistic picture than the theoretically derived functions, which are usually laborious to obtain and often approximate in nature. Other approaches to the calculation of molecular structures and poperties include semi-empirical SCF MO methods at various levels of sophistication and strictly ab initio

SCF MO methods. It should not be understood that the use of empirical and semi-empirical energy functions in strain energy calculations is "classical" in contrast to these latter quantum mechanical calculations. The basic difference may be that the calculation based on quantum mechanics is a deductive method, seeking to predict observable phenomena from the first principle (the Schrödinger equation), while the strain-energy minimization is an inductive method, searching a common analytical representation to a large assembly of observable phenomena. In fact, there is nothing classical in the energy functions, but they can be considered as an empirical representation of the Born-Oppenheimer approximation, according to which the ground state of a molecule is a continuous function of the atomic co-ordinates.

At an early stage of conformational energy calculations, the analysis was carried no further than the difference in strain energies between conformers of a single species with a fixed atomic arrangement. The use of digital computers made it possible to work out conformational energy surfaces of molecules or to find conformation of minimum potential energy. The more fundamental approach may be to try to obtain an analytical expression for potential functions that is applicable to a large group of molecules. This is known as the consistent force field (CCF) and was developed by Lifson and his co-workers (Lifson and Warshal, 1968; Lifson, 1972), to calculate helical structures of biopolymers and to correlate the physical properties of these systems with their conformation. For co-ordination compounds, Woldbye and his collaborators developed CCF calculations for a series of tris-bidentate complexes (Niketić and Woldbye, 1973a, 1973b, 1974; Niketić, Rasmussen, Woldbye and Lifson, 1976; Niketić and Rasmussen, 1977).

2 Conformational Energy of a Complex

The total conformational energy U of a complex can be expressed as a sum of five terms:

$$U = U_r + U_\theta + U_\phi + U_{nb} + U_{el} \tag{3.1}$$

where U_r is the potential energy for bond length distortion, U_θ, the potential energy for bond angle distortion, U_ϕ, the torsional potential energy, U_{nb}, non-bonded potential energy and finally U_{el}, electrostatic (Coulombic) interactions. In the literature this is described in various ways: total molecular potential energy, steric energy, strain energy etc. The total conformational energy thus defined is a measure of molecular strain within an isolated complex in a hypothetical state without any vibration. The absolute value of the total conformational energy has no intrinsic physical significance, since it depends upon the choice of the potential functions and energy parameters. The difference in the conformational energy for a series of conformers of the same kind of molecules is related to the molecular properties, which can be measured experimentally. Furthermore, the differences in strain energies provide a relative energy scale for a series of known conformers and also enable us to predict the stability of unknown conformers.

Table 3.1. Parameters for bond stretching potential functions

Bond	V_{ij} (kJ mol^{-1} A^{-2})	r_{ij}° (A)	Refs.
Co–N	1,205	2.00	a
N–C	3,616	1.47	b
C–C	3,014	1.54	b
C–H	3,014	1.093	b
N–H	3,399	1.011	a

a   Nakagawa and Shimanouchi, 1966.
b   Wiberg, 1965.

A. Bond Stretching Potential

In most of the conformational calculations the harmonic potential function is used:

$$U_r(r_{ij}) = \frac{1}{2} V_{ij}(r_{ij} - r_{ij}^{\circ})^2 \tag{3.2}$$

where r_{ij}° is the interatomic distance between the i-th and the j-th atom, r_{ij} is the equilibrium distance between them and V_{ij} is the force constant. Studies on infrared spectra and normal co-ordinate analyses of metal chelate compounds suggested that chelation does not affect the value of the bond stretching force constants of a ligand. Values of V_{ij}'s obtained from infrared and Raman spectra of the related compounds are listed in Table 3.1.

Table 3.2. Parameters for bond angle deformation potential

Angle	V_{θ} (kJ mol^{-1} deg^{-2})	θ_{ijk}° (deg)	Refs.
N–Co–N	198.0	90.0	a
Co–N–H	58.2	109.5	a
Co–N–C	116.4	109.5	a
N–C–C	324.4	109.5	b, c
N–C–H	190.9	109.5	b, c
H–N–H	160.3	109.5	d
C–N–H	190.9	109.5	d
H–C–H	160.3	109.5	b, c
H–C–C	190.9	109.5	b, c
C–C–C	324.4	109.5	b, c

a   Snow, 1970.
b   Wiberg, 1965.
c   Harris, 1966.
d   Niketić, 1974.

B. Potential Energy for Bond Angle Distortion

A number of analytical expressions have been proposed to account for the energies of the angle deformation modes in the infrared spectrum of a molecule. It is assumed that the small deformation in bond angles is subject to harmonic restoring forces and the simplest form of these expressions is:

$$V_\theta(\theta_{ijk}) = \tfrac{1}{2} V_{ijk}(\theta_{ijk} - \theta_{ijk}{}^\circ)^2, \tag{3.3}$$

where θ_{ijk} is a bond angle formed by three consecutive atoms i, j and k and $\theta_{ijk}{}^\circ$ is the unstrained angle.

The relevant force constants are listed in Table 3.2 together with the unstrained values.

C. Torsional Potential

Various spectroscopic, diffraction and thermodynamic measurements indicate that rotation around single bonds in polyatomic molecules is hindered by a potential energy barrier (Orville-Thomas, 1974). In 1937, Kemp and Pitzer assumed a torsional function

$$U_\phi = \tfrac{1}{2} V_\phi(1 + \cos 3\phi) \tag{3.4},$$

for the three-fold barrier in ethane and ϕ is the dihedral angle H–C–C–H. V_ϕ was calculated to be 13.2 kJ mol^{-1} on the basis of entropy and heat capacity data (Kemp and Pitzer, 1937). Torsional barriers determined from microwave spectra and thermodynamic data for a number of molecules containing C–C and C–N bonds with a three-fold or pseudo three-fold (as in CH_3NH_2) distribution of substituents revealed that the size of barriers around C–C bonds is fairly insensitive to the substituents and the height of the barrier is of the order of 13 ~17 kJ mol^{-1}.

The torsional energy around a bond between an octahedral six-co-ordinate metal and a ligand like ammonia or amine (three-fold or pseudo three-fold symmetry) is given by a sum of four such terms like Eq. (3.5) (Kim, 1960):

$$U_\phi = \sum_{i=1}^{4} \tfrac{1}{2} V_i'[1 + \cos 3\{\phi + (i-1) \times 90^\circ\}] \tag{3.5}$$

This term would approach zero if $V_1' = V_2' = \ldots = V_4'$. Moreover, Kim showed that the rotation of NH_3 groups in $[Co(NH_3)_6]^{3+}$ is free. Accordingly the torsional energy around the metal ligand bond in an octahedral complex is usually neglected (Gollogly and Hawkins, 1969).

D. Non-Bonded Interactions

It is generally accepted that interatomic forces exist other than those causing chemical bond formation. Molecules assemble together to form liquids and solids, for example, by some kind of attractive forces, however, when they come close together they repel each other by strong repulsive forces and they are held apart. Similar interactions are considered to exist between nonbonded atoms in a molecule. These interactions are often very important in determining the molecular configurations and conformations. Though theoretically not justifiable, intermolecular potential functions are widely used in order to treat the intramolecular non-bonded interactions. Figure 3.1 shows an example of general behaviour of interatomic potential with interatomic distance.

The attractive forces at larger separations arise from the coupling of electric dipoles, one being a dipole fluctuation in one molecule and the other the dipole induced by it in the other. London developed the quantum mechanical theory of these interactions (London, 1930, 1937) and the forces are known as London dispersion forces. He showed that the attractive potential depends on the polarisabilities and ionisation potentials of each attracting atom and its dependence on distance is as r^{-6} (Pitzer, 1959).

When the two atoms come close together and the electron clouds of the two atoms begin to overlap, the repulsive force becomes predominant. The expression for the repulsive potential is given by an exponential form or inverse power of r. The exponential form of the repulsive potential is theoretically more justifiable. Since the form of the wave function is exponential, it is logical that the interpenetration (repulsive) energy should vary exponentially (Heitler and London, 1927).

Various types of expression have been proposed for non-bonded potential functions. The most general form between a pair of atoms i and j is:

$$U(r_{ij})_{\text{nb}} = \frac{a_{ij} \exp(-b_{ij}r_{ij})}{r_{ij}{}^{d_{ij}}} - \frac{c_{ij}}{r_{ij}^6} \tag{3.6}$$

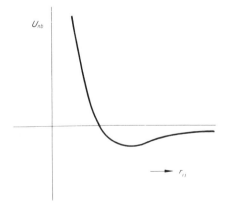

Fig. 3.1. Typical potential energy curve for non-bonded interactions

where the first term represents the repulsive potential and the second the attractive potential. When $b_{ij} = 0$ and $d_{ij} = 12$, the function becomes

$$U(r_{ij})_{nb} = \frac{a_{ij}}{r_{ij}^{12}} - \frac{c_{ij}}{r_{ij}^{6}} \qquad (3.7)$$

This is called the Lennard-Jones potential which was suggested on purely empirical grounds (Lennard-Jones, 1929). And if $d_{ij} = 0$, the function reduces to the Buckingham type, i.e.,

$$U(r_{ij})_{nb} = a_{ij} \exp\left(-b_{ij}r_{ij}\right) - c_{ij}/r_{ij}^{6} \qquad (3.8)$$

A number of other interatomic potential functions have been proposed (see for example, Hirschfelder, Curtis and Bird, 1954; Torrens, 1972). The constants in these equations are determined by comparing various calculated potential energy curves with experimental ones for gaseous molecules like methane.

For the non-bonded H...H distance of most interest to the study of chelate rings, ranging from 2.0 Å to 3.0 Å, Mason and Kreevoy's values of a_{ij}, b_{ij} and c_{ij} in Eq. (3.8) give rise to relatively large energies, namely the potential function is "hard" (1955). On the other hand, Hill's parameters make the equation "soft" and it yields very small interaction energies (1948). Accordingly the coefficients are selected in such a way that the resulting potential curve lies somewhere between the "hard" and "soft" curves. As an example, Table 3.3 presents two sets of coefficients, I, and II, which were used for conformational analysis of tris-diamine complexes with six-membered chelate rings (Niketić and Woldbye, 1973). The first set given by Liquori and coworkers is derived from the experimental data of DeCoen and co-workers (Liquori, Damiani and Elefante, 1968; DeCoen, Elefante, Liquori and Damiani, 1967). The second set given by Ramachandran and co-workers gives rather "soft" H...H functions (Ramachandran, Venkatachalam and Krimm, 1966). However, the functions for other interactions are harder than the first set. The first set is widely used for a number of co-ordination compounds.

Table 3.3. Parameters for non-bonded potential functions (in kJ/atom pair)

	I			II		
	$a_{ij} \times 10^{-4}$	b_{ij} (Å^{-1})	c_{ij} (Å^6)	$a_{ij} \times 10^{-4}$	b_{ij}	c_{ij}
H...H	2.76	4.08	205.9	3.47	4.6	195.9
H...C	13.14	4.20	506.9	32.61	4.6	694.0
H...N	11.76	4.32	415.2	22.35	4.6	652.9
C...C	99.20	4.32	1,246.4	386.7	4.6	2,510.9
C...N	88.77	4.44	1,021.3	253.2	4.6	2,390.8
N...N	78.02	4.55	837.1	169.1	4.6	2,289.0

E. Electrostatic Interactions

If point charges, permanent dipoles or higher multipoles exist in a molecule, their Coulombic interaction must be taken into account. A straightforward model uses a simple Coulombic potential function between two partial charges, q_i and q_j separated by a distance r_{ij}:

$$V_{el}(r_{ij}) = - \Sigma \, q_i q_j / D \, r_{ij} \tag{3.9},$$

where D is the effective dielectric constant.

The charges can be estimated from the observed accurate electron density in crystals (cf. Chapter V). The charges are mostly assigned so that the observed dipole moments or the values of bond moments can be reasonably reproduced.

3 Computational Methods for Conformational Analysis

The total conformational energy of a molecule can be written down as a function of the n parameters x_i ($i = 1, 2, \ldots n$) that define the molecular geometry.

$$V = V(x_1, x_2, x_3 \ldots x_i, \ldots x_n) \tag{3.10}$$

This general equation forms an n dimensional conformational energy surface. The minima correspond to various stable conformations, whereas the saddle points (maxima) between them represent the activated complexes for the change from one stable conformation to the other.

Most minimization methods involve selection of a set of trial structures and minimizing their energy by iterative procedures. The computer program for solving the problem starts from a trial structure and moves down the energy surface along the steepest descent path (Wiberg, 1965; Bixon and Lifson, 1967). First the gradients of the potential surface at the point corresponding to an assumed trial structure are evaluated for searching the nearest energy minima:

$$(\partial V / \partial x_1, \; \partial V / \partial x_2, \ldots \partial V / \partial x_i, \ldots \partial V / \partial x_n).$$

They define the search directions: the increment δx_i in the search direction $\partial V / \partial x_i$ is obtained by one dimensional minimization of $V(x_i - \partial V / \partial x_i \cdot \delta x_i)$, i.e.

$$\frac{d}{dx_i} \, V\!\left(x_i - \frac{\partial V}{\partial x_i} \, \delta x_i \right) = 0 \tag{3.11}$$

in other words, δx_i is varied until the energy V stops decreasing. Using δx_i values a new set of parameters x_i's are obtained which may be used for recalculation of the second set of δx_i's. The procedure may be repeated until δx_i becomes zero. The basic ideas and the advantages of the method of the steepest descent have been discussed (Gleicher and Schleyer, 1967; Schleyer, 1971; Allinger, Hirsch, Miller, Tyminski and VanCatledge, 1968; Allinger, Tribble, Miller and Werz, 1971; Williams, Stang and Schleyer, 1968).

The convergence of this method of the steepest descent is rapid when x_i's are far from minimum, but too slow near the minimum. At this stage the modified Newton method is used. A set of equations for δx_i is obtained as follows:

V at the minimum energy point, $(x_1^\circ, x_2^\circ, \ldots x_n^\circ)$ is expanded around a point near it, $(x_1, x_2, \ldots x_n)$ to the second order:

$$V(x_1^\circ, x_2^\circ, \ldots x_n^\circ) = V(x_1, x_2, \ldots x_n) + \sum_i \frac{\partial V}{\partial x_i} \delta x_i + \frac{1}{2} \sum_i \sum_j \frac{\partial^2 V}{\partial x_i \partial x_j} \delta x_i \delta x_j \qquad (3.12)$$

The necessary condition that $V(x_1^\circ, x_2^\circ, \ldots x_n^\circ)$ is minimum requires:

$$\frac{\partial V(x_1^\circ, x_2^\circ, \ldots x_n^\circ)}{\partial x_i} = 0, \quad i = 1, 2, \ldots, n \qquad (3.13)$$

Eqs. (3.12) (to the first order) and (3.13) lead to a set of linear equations:

$$\sum_j \frac{\partial^2}{\partial x_i \partial x_j} V(x_1, x_2, \ldots x_n) \delta x_j = - \frac{\partial V(x_1, x_2, \ldots x_n)}{\partial x_i} \quad i = 1, 2, \ldots n \qquad (3.14)$$

δx_j's are obtained by a set of linear equations (3.14). The iterative procedure is repeated as before.

A starting set of molecular parameters can be obtained from the results of X-ray crystal structure analysis. If the positions of hydrogen atoms bonded to carbon and nitrogen atoms are not available from X-ray data, they are located by calculation assuming tetrahedral angles and appropriate bond distances. If the structure is unknown, the model may be constructed on the basis of known interatomic distances and bond angles. The iteration is terminated and the molecular geometry is calculated from the final set of parameters when the parameter shifts became less than twice the observed standard deviations of bond distances or bond angles, say 0.01 Å and 5°. The bond lengths and angles in the complexes can be reproduced to within several times the standard deviations of the values obtained by crystal structure analysis.

4 Geometrical Molecular Models

The results of strain energy minimization emphasize that regular models (for instance, Dreiding), although useful for qualitative analysis, are at best crude for any quantitative evaluation of relative stabilities. In setting up models, the rod-and-cylinder method of linking sometimes cause an irregular distortion of a few bond angles of a complicated multidentate structure, since the rods are more flexible than the cylinders. When the more flexible rods are used (for example rods made of plastic) for constructing multidentate complexes, the angles are found to be almost identical to those determined by means of X-rays.

5 Conclusion

As will be shown in the next Chapter for particular complexes, studies on various co-ordination compounds revealed that energy minimization calculations can predict the detailed geometries of the complexes. Where particular contacts are absent, the computed geometry of a co-ordination compound agrees with that observed in the crystal structure to within several standard deviations. Relative stabilities of various conformers can also be successfully predicted with reasonable certainty, unless other factors are operative. The calculation indicated that angular deformations are generally important in deciding the relative stabilities of the conformers. The bond angles deform with comparatively small expenditure of energy for quite large angular changes. Likewise torsional distortions occur easily and both deformations alleviate close non-bonded interactions. A close balance is often observed between torsional and angular distortions against non-bonded terms.

Chapter IV Structure and Isomerism
of Optically Active Complexes

Most of the optically active complexes for which experimental and theoretical study has been made involve chelate rings. Thus structure and isomerism of metal chelate complexes will be discussed in this chapter. One of the most striking properties of chelate compounds is their unusual stability due to chelate ring formation. Structures of four-, five-, six- and seven-membered chelate rings have been reported, even eight- and larger membered chelate rings are known. Bonds in chelate rings may arise from two general types of groups: i) primary acid groups in which a metal ion replaces an acid hydrogen and ii) neutral groups which contain an atom with a lone pair suitable for bond-formation. When five-membered chelate rings are formed the resulting complex is most stable.

1 Bidentates

A. Four-Membered Chelate Rings

The stereochemistry of metal chelate rings differs from that of carbon ring systems in that all of the atoms in the ring are not the same size and some of the bond angles normally vary from $109.5°$ (or $120°$) as a result of the directed valences of the metal ion. These two factors may alleviate the strain involved in the four-membered ring systems. As an example the molecular geometry of the carbonatocobalt (III) ring

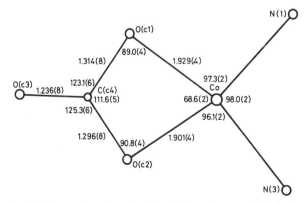

Fig. 4.1. Bond lengths and angles of the carbonatocobalt (III) system (Toriumi and Saito, 1975)

in $(+)_{589}cis$-β-$[Co(CO_3)(3(S)8(S)$-$2'$, $2,2'$-tet)$]ClO_4$ is presented in Fig. 4.1 (Toriumi and Saito, 1975). The carbonato group is planar but makes an angle of 4.3° with the plane formed by Co, N(1) and N(3). The OCoO angle is compressed to 68.6(2)°, which is 90° in a strain free state. The dimensions of the co-ordinated carbonato group closely resemble those of other reported cobalt-carbonato systems (Kaas and Sørensen, 1973; Geue and Snow, 1971; Barclay and Hoskins, 1962).

Other examples of bidentate ligands capable of forming four-membered chelate rings are: nitrate, sulphate, sulphite and xanthate etc.

B. Five-Membered Chelate Rings

As early as 1933 Rosenblatt and Schleede suggested in their paper on platinum complexes that the five-membered chelate ring formed by ethylenediamine need not be planar. Six years later Kobayashi studied the circular dichroism of $(+)_{589}[Co(en)_3]Br_3$ and made the same suggestion (1939). This was indeed verified by a crystal structure analysis of $[Cu(en)_2][Hg(NCS)_4]$ (Scouloudi, 1950) and then of $[CoCl_2(en)_2]Cl \cdot HCl \cdot 2H_2O$ by Nakahara, Saito and Kuroya (1952).

Since the five-membered chelate rings formed by ethylenediamine are puckered, there exist two enantiomeric conformations δ and λ, as shown in Fig. 4.2.

In the $trans$-dihalogenobis(ethylenediamine)cobalt(III) ion, the following three combinations of two chelate rings are possible: $(\delta\delta)$, $(\lambda\lambda)$ and $(\delta\lambda)$. $(\delta\delta)$ and $(\lambda\lambda)$ are optical isomers and have the same relative strain energy. The relative strain energies of $(\delta\delta)$ and $(\delta\lambda)$, must however, be different. This energy difference has been calculated by Corey and Bailar who found that the $(\delta\delta)$ configuration is more stable by about 4.2 kJ mole^{-1} (1959). In this calculation only the interaction between non-bonded hydrogen atoms was considered for rigid structures. In crystals of $trans$-$[CoX_2(en)_2]X \cdot HX \cdot 2H_2O$ (X = Cl, Br), the complex ion possesses a centre of symmetry, thus being $(\delta\lambda)$ form (Nakahara, Saito and Kuroya, 1952; Ooi, Komiyama, Saito and Kuroya, 1959). The $(\delta\lambda)$ form must be favoured by specific intermolecular forces in the crystalline state, such as hydrogen bonding.

When the bidentate is propylenediamine, in which a methylene hydrogen atom is replaced by a methyl group, the stereochemistry is further complicated, since the ligand itself is optically active. From an X-ray examination of $trans$-$[CoCl_2(pn)_2]Cl \cdot HCl \cdot 2H_2O$, obtained from rac-pn as a starting material, it was revealed that the

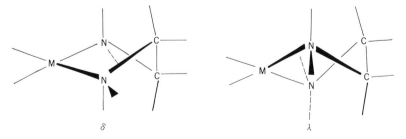

Fig. 4.2. Two possible conformations of a metal-ethylenediamine ring

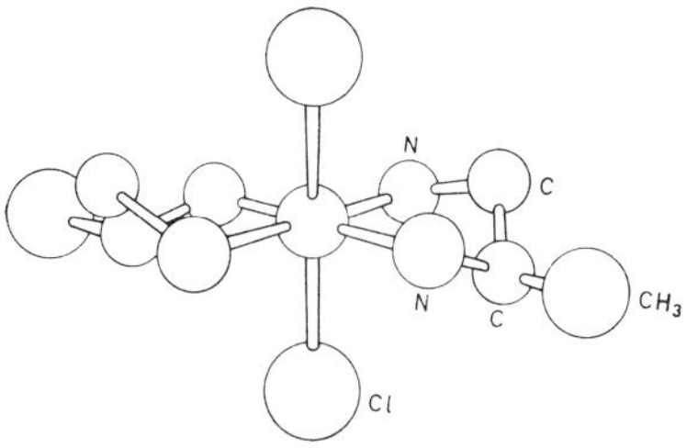

Fig. 4.3. $(-)_{589}$*trans*-$[CoCl_2(R-pn)_2]^+$ (Saito and Iwasaki, 1962)[9]

general features of the crystal structure resemble those of the ethylenediamine analogue and the complex ion is again centro-symmetric and assumes the $(\delta\lambda)$ form,
namely *trans*-$[CoCl_2(R-pn)(S-pn)]^+$. But, if optically active propylenediamine is
used, the resulting complex ion is no longer centrosymmetric. There are six possible
stereoisomers of the *trans*-$[CoX_2(R-pn)_2]^+$ ion. Firstly, there are two possible directions of each C—CH$_3$ bond with respect to the chelate ring. The C—CH$_3$ bond can lie
approximately parallel to the "average" plane of the five-membered chelate ring or it
can stand approximately perpendicular to the plane of the chelate ring. These forms
are called "equatorial" and "axial" respectively. Secondly, there are two possible dis-
positions of the methyl groups attached to the two chelate rings, namely *trans* and
cis isomers. Thus six possible isomers may be represented as follows:

cis(ax, ax), *cis*(ax, eq), *cis*(eq, eq),
trans(ax, ax), *trans*(ax, eq), *trans*(eq, eq).

The *trans*-$[CoCl_2(R-pn)_2]^+$ has been shown to have the *trans*(eq, eq) form (Fig. 4.3)
(Saito and Iwasaki, 1962). In fact, the methyl group cannot have axial disposition
owing to the repulsion between the bulky methyl group and the chlorine atom. The
complex ion has an approximately two-fold axis along the Cl—Co—Cl bond. The
conformation of the chelate ring is λ. The absolute configuration of the $R(-)$-propyl-
enediamine molecule is represented by the formula:

$$\begin{array}{c} CH_2NH_2 \\ | \\ H\text{---}C\text{---}NH_2 \\ | \\ CH_3 \end{array}$$

This is in agreement with the result determined chemically with reference to $R(-)$-
alanine (Reihlen, Weinbrenner and Hessling, 1932). It is to be noted here that when a
molecule of R-propylenediamine is co-ordinated to a metal atom to form a chelate
ring with its C—CH$_3$ bond in the equatorial position, the conformation of the chelate
ring necessarily becomes λ. The equatorial preference of the methyl group is the
same as in methylcyclohexane, however, the equatorial preference of the methyl

---

9   All the figures in this Chapter correctly represent the absolute configuration.

group is much greater in the metal complex, since there are strong steric interactions in the complex with other ligands as mentioned above. In contrast to this there are only 1,3 interactions with the hydrogen atoms in the case of cyclohexane. Substitution on the ligating nitrogen atoms produces similar results. In N-methylethylenediamine the difference between the axial and equatorial position is less pronounced than the substitution on methylene groups. The analogous complex ion, $(-)_{589}$-$[CoCl_2(N\text{-meen})_2]^+$ has a two-fold axis through the Cl–Co–Cl bond and the conformation of the two five-membered chelate rings is δ. Complexes of this type have stereochemical interest since the co-ordinated nitrogen centre is dissymmetric. The absolute configuration of the two nitrogen atoms is R. The two methyl groups are in equatorial positions (Robinson, Buckingham, Chandler, Marzilli and Sargeson, 1969). The structures of the complex ions, $trans$-$[PtX_2(-\text{chxn})_2]^{2-}$ (X = Cl, Br) and $[Pt(-\text{chxn})_2]^{2+}$ are known (Larsen and Toftlund, 1977). The geometry of the five-membered chelate rings is similar to that of the Co-en ring, the conformation being λ. The fused cyclohexane ring takes a chair conformation.

When three bidentate ligands are co-ordinated octahedrally to a central metal atom, two optical isomers Λ and Δ can occur (XVIII and XIX in Chapter I). This type of isomerism has been proved for a number of compounds. The first investigation by X-ray diffraction was that of tris(oxalato)chromate(III) ion (van Niekerk and Schoening, 1952). The absolute configuration of the complex ion was determined by Butler and Snow (1971). $(+)_{589}[Cr(ox)_3]^{3-}$ has the absolute configuration Λ.

Ethylenediamine

Unlike the planar Cr-ox ring, the Co-en ring is puckered and dissymmetric. There are eight possible configurations for tris(ethylenediamine)cobalt(III) ions, viz.

$$1\begin{vmatrix} \Lambda(\delta\delta\delta) \\ \Delta(\lambda\lambda\lambda) \end{vmatrix} \quad 2\begin{vmatrix} \Lambda(\delta\delta\lambda) \\ \Delta(\lambda\lambda\delta) \end{vmatrix} \quad 3\begin{vmatrix} \Lambda(\delta\lambda\lambda) \\ \Delta(\lambda\delta\delta) \end{vmatrix} \quad 4\begin{vmatrix} \Lambda(\lambda\lambda\lambda) \\ \Delta(\delta\delta\delta) \end{vmatrix}$$

They form two catoptric series and Fig. 4.4 shows the four diastereoisomers of the Λ series. The C–C axis is eclipsed in the combination, $\Lambda(\delta)$ and staggered in the combination, $\Lambda(\lambda)$. In the former the C–C axis is nearly parallel to the pseudo-three-fold axis of the complex ion, while it is largely slanted obliquely in the latter. Accordingly they are called lel and ob conformations respectively. Thus the four diastereoisomers can be designated as lel_3-, lel_2ob-, $lelob_2$- and ob_3-isomers (Corey and Bailar, 1959).

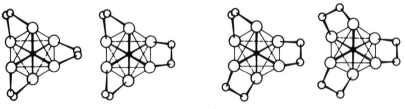

Fig. 4.4. Four diastereoisomers of Λ-$[Co(en)_3]^{3+}$

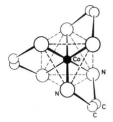

Fig. 4.5. $\Lambda(+)_{589}[\text{Co(en)}_3{}^{\delta\delta\delta}]^{3+}$

Figure 4.5 shows a perspective drawing of the $(+)_{589}$ tris(ethylenediamine)cobalt-(III) complex ion. This is the first chelate complex whose absolute configuration was determined by means of X-rays (Saito, Nakatsu, Shiro and Kuroya, 1955). The absolute configuration is the $\Lambda(\delta\delta\delta)$, lel_3 form, in agreement with the result of calculation that the lel_3 form is more stable by about 7.5 kJ mol^{-1} than the ob_3 form (Corey and Bailar, 1959). Several crystal structures containing this type of complex ion have been known: $(+)_{589}[\text{Co(en)}_3]_2\text{Cl}_6 \cdot \text{NaCl} \cdot 6\text{H}_2\text{O}$ (Saito, Nakatsu, Shiro and Kuroya, 1955, 1957); $(+)_{589}[\text{Co(en)}_3]\text{Br}_3 \cdot \text{H}_2\text{O}$ (Nakatsu, 1962); $(+)_{589}[\text{Co(en)}_3]$-$\text{Cl}_3 \cdot \text{H}_2\text{O}$ (Iwata, Nakatsu and Saito, 1969); $(+)_{589}[\text{Co(en)}_3](\text{NO}_3)_3$ (Witiak, Clardy and Martin, 1972); $[\text{Co(en)}_3]_2(\text{HPO}_4)_3 \cdot 9\text{H}_2\text{O}$ (Duesler and Raymond, 1971). The complex ion has D_3 symmetry within the limits of experimental error. The shape and the size of the cobalt-ethylenediamine ring are as follows:

Co–N = 1.978 ± 0.004 Å	NCoN = 85.4° ± 0.3°
N–C = 1.497 ± 0.010 Å	CoNC = 108.4° ± 0.5°
C–C = 1.510 ± 0.010 Å	NCC = 105.8° ± 0.7°
	N–C–C–N = 55.0°* * dihedral angle

The C–C bond length in the chelate ring is significantly shorter than the normal C–C bond length of 1.544 Å (diamond). A neutron diffraction study of $[\text{Co}(-\text{pn})_3]\text{Br}_3$ gave the C–C bond length of 1.518 Å (Shintani, Sato and Saito, 1979). The octahedron formed by the six nitrogen atoms is trigonally twisted and slightly compressed along the three-fold axis: the upper triangle formed by the three nitrogen atoms is rotated counterclockwise by about 5° with respect to the lower triangle formed by the remaining three nitrogen atoms from the position expected for a regular octahedron. The Co–N bond makes an angle of 55.9° with respect to the three-fold axis of the complex ion. This angle is 54.75° for a regular octahedron.

The carbon atoms of the chealte ring show thermal anisotropy best described as an oscillatory motion perpendicular to the C–C bond, $(+)_{589}[\text{Co(en)}_3]\text{Cl}_3 \cdot \text{H}_2\text{O}$, $[\text{Co(NCS)(SO}_3)(\text{en})_2]$ (Baggio and Becka, 1969); $[\text{Cu(en)}_2(\text{H}_2\text{O})]^{2+}$ (Williams, Larsen and Cromer, 1972); $cis\text{-}[\text{Co(NO}_2)_2(\text{en})_2]^+$ (Shintani, Sato and Saito, 1976). Hawkins calculated the conformational energy of a five-membered metal-ethylenediamine ring (1971). The calculation was carried out by changing structural parameters defining the conformation of the chelate ring. It was revealed that a whole range of conformations of minimum strain energy could be interrelated by holding the cobalt and nitrogen atoms steady while flapping the two carbon atoms on both sides of the plane formed by the cobalt and the two nitrogen atoms, maintaining a relatively constant dihedral angle N–C–C–N. This equality in energy between various symmetric and asymmetric skew conformations was supported by the observed conformations of the chelate rings in various crystals. The observed feature of anisotropic vibrations of the carbon atoms mentioned above seems to support the existence of a puckering motion of the chelate ring in solution. Actually Mason and Norman (1964, 1965) measured the circular dichroism spectra of $[Co(en)_3]^{3+}$ ions in solution and suggested that different conformations of $[Co(en)_3]^{3+}$ coexist in solution (McCaffery, Mason and Norman, 1965a). Beattie examined the nmr spectra of $[Co(en)_3]^{3+}$ and showed that the ligands undergo rapid inversion between δ and λ conformations in solution (1971). Considering the statistical effect he suggested that the most abundant conformation in solution may be $\Lambda(\delta\delta\lambda)$ and not $\Lambda(\delta\delta\delta)$.

Recently the complex ions, $[Co(en)_3]^{3+}$ with $\Lambda(\delta\delta\lambda)$ and $\Delta(\lambda\lambda\delta)$ configurations were recognized in crystals of $[Co(en)_3][SnCl_3]Cl_2$ (Haupt, Huber and Preut, 1976). The lel_2ob conformation of the complex ion favoured N–H ... Cl hydrogen-bond formation in crystals, resulting in the stabilization of the lel_2ob conformation.

The complex ion, $[Cr(en)_3]^{3+}$ takes the lel_3 form in crystals of racemic $[Cr(en)_3]$-$Cl_3 \cdot 3H_2O$ (Whuler, Brouty and Herpin, 1975). All the other three possible conformations of $[Cr(en)_3]^{3+}$ were reported for the first time by Ibers and his co-workers in 1968. Previously all the crystal structures containing the $[M(en)_3]^{3+}$ complex ion were found to hold the lel_3 conformers. In $[Cr(en)_3][Ni(CN)_5] \cdot 1.5H_2O$ the complex cation takes lel_2ob and $lelob_2$ conformations (Raymond, Corfield and Ibers, 1968), whereas it takes the ob_3 conformation in $[Cr(en)_3][Co(CN)_6] \cdot 6H_2O$ (Raymond and Ibers, 1968). It turned out that hydrogen bonding specifically stabilises these conformations other than lel_3 since the crystal structure permits more hydrogen bonds than the lel_3 form. The lel_3 form is the most compact and probably leads to better packing in the lattice. In crystals of rac-$[Cr(en)_3](SCN)_3 \cdot 0.75H_2O$ at 293 K, the conformation of the chelate rings is disordered. For the complex ion, Λ-$[Cr(en)_3]^{3+}$, one of the chelate rings exhibits disorder and the conformation can be represented as $\Lambda[\delta\delta(0.70 \, \delta + 0.30 \, \lambda)]$. On lowering the temperature to 133 K, the conformation changes to $\Lambda(\delta\delta\delta)$, whereby the unit cell volume contracts by 7.2%, reflecting the more compact lel_3 conformers, but no change is observed in the packing modes (Brouty, Spinat, Whuler and Herpin, 1977). Thus the lel_3 conformation of $[Cr(en)_3]^{3+}$ appears to realize less frequently at ambient temperatures than that of the Co analogue.

Recently Niketić and Rasmussen (1978) used a fast convergent energy minimisation programme to calculate equilibrium conformations of the $[Me(en)_3]$ system. Table 4.1 lists various energy contributions. As can be seen from the Table, the ob_3 isomer is less stable by 4.75 kJ mol^{-1} in agreement with Corey and Bailar's classical calculation (1959). The average ob-lel energy difference is 1.6 ± 0.9 kJ mol^{-1}. Their detailed analysis of the $[Me(en)_3]$ system revealed the following features for the equilibrium conformation. Firstly, when the five-membered chelate ring carries no methyl group, the ring is highly puckered, the mean dihedral angle N–C–C–N being 55.3 ± 1.5°. Secondly, a $[CoN_6]$ octahedron is twisted around the three-fold axis, the upper triangle formed by three nitrogen atoms is rotated by about 5° from the position expected for a regular octahedron. The octahedron is

Table 4.1. Energy contributions for the [Me(en)$_3$] system

	lel_3	lel_2ob	$lelob_2$	ob_3
Bond stretching deformations	1.11	1.26	1.39	1.34
Bond angle deformations	8.84	8.38	8.24	8.08
Torsional strain	16.89	18.21	19.11	19.44
Non-bonded interactions	−19.68	−18.01	−16.81	−16.96
Total conformational energy	7.16	9.84	11.93	11.91
Difference	0.00	2.68	4.77	4.75

All energies are given in kJ mol^{-1}.

slightly compressed along the three-fold axis. These results agree satisfactorily with the observed geometries of [Co(en)$_3$]$^{3+}$ and [Cr(en)$_3$]$^{3+}$ in actual crystal structures as described earlier (see also p. 57).

Propylenediamine

When the bidentate is propylenediamine the number of isomers for the [Co-(±pn)$_3$]$^{3+}$ system increases to 24, where equatorial preference of the C–CH$_3$ bond is assumed (Harnung, Kallesøe, Sargeson and Schäffer, 1974). They constitute two catoptric series with absolute configuration Λ and Δ respectively. In each of the series there exist two types of geometric isomers, *fac* and *mer*, for lel_3 and ob_3 conformers and four with lel_2ob and $lelob_2$ isomers. They are tabulated in Table 4.2. Jaeger and Blumendal supported absolute stereospecificity of optically active ligands, thus it was believed that (−)-propylenediamine favoured the formation of only the (−)$_{589}$[Co(−pn)$_3$]$^{3+}$ isomer to the complete exclusion of the other. In 1959, Dwyer

Table 4.2. Isomers of [Co(±pn)$_3$]$^{3+}$

Configuration	Optical isomers	Geometrical isomers relating to methyl group	Number of isomers
lel_3	Λ (+++)	*fac, mer*	2
	Δ (−−−)	*fac, mer*	2
lel_2ob	Λ (++−)	*fac, mer*(3)	4
	Δ (−−+)	*fac, mer*(3)	4
$lelob_2$	Λ (++−)	*fac, mer*(3)	4
	Δ (−−+)	*fac, mer*(3)	4
ob_3	Λ (−−−)	*fac, mer*	2
	Δ (+++)	*fac, mer*	2
		Total	24

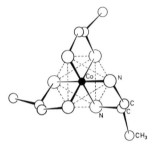

Fig. 4.6. $(-)_{589}[Co(-pn)_3]^{3+}$(Iwasaki and Saito, 1966)

and his collaborators first succeeded in separating the isomers of $[Co(-pn)_3]^{3+}$ by fractional crystallisation (Dwyer and Garvan, 1959) and by cellulose column chromatography (Dwyer, Sargeson and James 1964). They isolated $(+)_{589}$- and $(-)_{589}[Co(-pn)_3]^{3+}$. MacDermott (1968) separated the $\Delta(lel_3)(mer)$ isomer from the $\Delta(lel_3)$-(fac) isomer by fractional crystallisation.

Yamasaki and his collaborators isolated $\Lambda(ob_3)(fac)$ and $\Lambda(ob_3)(mer)$ isomers of $[Co(-pn)_3]^{3+}$ by column chromatography on an ion exchange SP-sephadex (Kojima, Yoshikawa and Yamasaki, 1973). Among the 24 isomers of $[Co(\pm pn)_3]^{3+}$, the structures of only two facial isomers are known in detail. Figure 4.6 illustrates the most stable isomer, $(-)_{589}[Co(-pn)_3]^{3+}$ (Iwasaki and Saito, 1966). The absolute configuration of the complex ion can be designated the $\Delta(\lambda\lambda\lambda)$, lel_3 form. The three methyl groups are attached in facial positions. The geometry of the five-membered chelate ring resembles that of the cobalt-ethylenediamine ring. Methyl substitution on the chelate ring does not seem to disturb the overall features of the rings, with the $C-CH_3$ bond in the equatorial position. Methyl groups were shown to take two alternative azimuthal orientations around the $C-CH_3$ bond (Kuroda, Shimanouchi and Saito, 1975).

Figure 4.7 shows a perspective drawing of the complex ion, $(+)_{589}[Co(-pn)_3]^{3+}$, as viewed down the pseudo-three-fold axis (Kuroda and Saito, 1974). This is the facial isomer. The three chelate rings take the ob conformation with methyl groups in equatorial positions. The absolute configuration can be designated as $\Lambda(\lambda\lambda\lambda)$. The geometry of the chelate ring is not very different from that of the lel_3 fac isomer. The only difference is that the $N-Co-N$ angle is compressed by about $1.7°$.

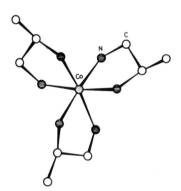

Fig. 4.7. $(+)_{589}[Co(-pn)_3]^{3+}$ (Kuroda and Saito, 1974)

In contrast to the facial isomers, it has not yet been possible to determine the structures of any *mer* isomers since the salts containing *mer* isomers are amorphous glasses or the complex ions exhibit orientational disorder in the crystal lattice. For example, *mer*-$(-)_{589}[Co(-pn)_3](+)_{589}[Cr(mal)_3] \cdot 3H_2O$ crystallizes in a rhombohedral space group, $R32$ and the complex cations are on a set of special positions with D_3 site symmetry. *mer*-$[Co(-pn)_3]^{3+}$ has no strict overall symmetry, but the non-methylated fragment does have D_3 symmetry. The electron-density distribution of the complex cation in the crystal looks like that of the tris $(R,R$-2,3-diaminobutane)cobalt(III) ion with a methyl group of a half weight (Butler and Snow, 1971, 1976). The hexacyanocobaltate(III) salt of the $\Lambda(ob_3)(mer)$ isomer is cubic. Again the complex ions exhibit orientational disorder and no information was obtained on the structure of the complex cation (Kuroda and Saito, unpublished work). *mer*-$(-)_{589}[Co(-pn)_3]Br_3 \cdot 2H_2O$ is an amorphous glass (Crossing and Snow, 1972). It appears that the *mer* isomer can interfere with the crystal packing and amorphous glass or disordered orientation on the regular crystal lattice result with a structure which grossly resembles that of the facial isomer.

The free energy differences at 298 K of the lel_3 and ob_3 isomers were calculated from the equilibrium concentration of isomers (Dwyer, MacDermott and Sargeson, 1963). The lel_3 isomer is more stable by 6.7 kJ mol^{-1}. Recently Schäffer and his collaborators determined the free energy differences between the isomers at 373 K as follows (Harnung, Kallesøe, Sargeson and Schäffer, 1974):

$$\Delta G° \ (ob_3 \to lel_3) = -6.73 \text{ kJ mol}^{-1}$$
$$\Delta G° \ (ob_2 lel \to lel_2 ob) = -2.56$$
$$\Delta G° \ (lel_2 ob \to lel_3) = +0.50$$

Calculation of the conformational energy showed that the presence of a methyl group in an equatorial position changes the energy by -1.53 kJ mol^{-1}, and in an axial position by 13.53 kJ mol^{-1} (Niketić and Rasmussen, 1978).

trans-1,2-Diaminocyclohexane

The existence of a fused cyclohexane ring fixes the puckering motion of the five-membered chelate ring formed by *trans*-1,2-diaminocyclohexane. The isomers of $[Co(\pm chxn)_3]^{3+}$ comprise two catoptric series with absolute configurations Δ and Λ around the cobalt atom. The absolute configurations of the chelate rings formed by $(-)R,R$-chxn and $(+)S,S$-chxn are λ and δ, respectively. For each of the configurational series, there exists four diastereoisomers: lel_3, $lel_2 ob$, $ob_2 lel$ and ob_3 (Harnung, Sørensen, Creaser, Malgaard, Pfenninger and Schäffer, 1976). In Fig. 4.8 a perspective drawing is shown of the complex ion, $(-)_{589}[Co(+chxn)_3]^{3+}$. The complex ion has an approximate D_3 symmetry. The absolute configuration is Λ-$(\delta\delta\delta)$, lel_3 (Marumo, Utsumi and Saito, 1970; Miyamae, Sato and Saito, 1979). The geometry of the chelate ring system is very much like that of $[Co(en)_3]^{3+}$. The average dihedral angle of N–C–C–N is 59.3°, which is almost identical with the

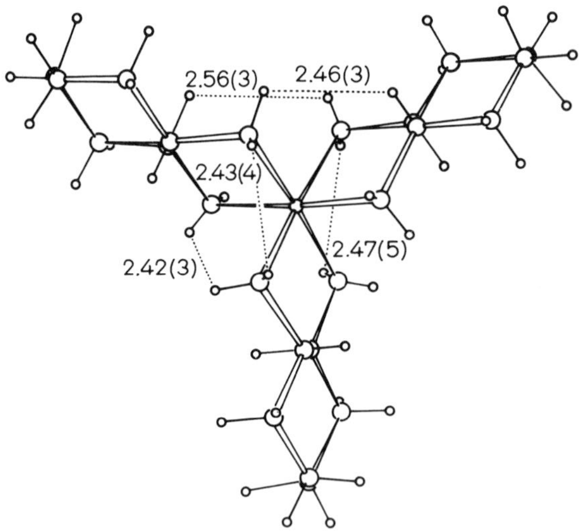

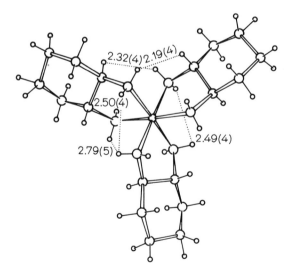

Fig. 4.8. $(-)_{589}[\mathrm{Co}(+\mathrm{chxn})_3]^{3+}$ (Miyamae, Sato and Saito, 1979). Non-bonded H...H interactions are indicated by broken lines

value expected in a free ligand molecule. The cyclohexane ring takes a chair conformation and its shape and size indicate no strain in the ring system.

In the ob_3 isomer, $(+)_{589}[\mathrm{Co}(-\mathrm{chxn})_3]^{3+}$, shown in Fig. 4.9, the central C–C bond in the chelate ring is inclined at an angle of 66° with respect to the three-fold axis, whereas it was nearly parallel to the three-fold axis in the case of the lel_3 isomer. The geometry of the Co-chxn ring is similar to that in the lel_3 isomer, however, the NCoN angle is 84.2°, being compressed by about 2.4° compared to the lel_3 isomer. The absolute configuration is $\Lambda(\lambda\lambda\lambda)$ (Kobayashi, Marumo and Saito, 1972a). Non-bonded H ... H interactions shown in Figs. 4.8 and 4.9 clearly indicate that the lel_3 isomer is more stable than the ob_3 isomer.

Fig. 4.9. $(+)_{589}[\mathrm{Co}(-\mathrm{chxn})_3]^{3+}$ (Miyamae, Sato and Saito, 1979). Non-bonded H...H interactions are indicated by broken lines

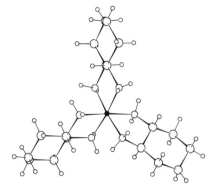

Fig. 4.10. $(-)_{589}[Co(+chxn)_2(-chxn)]^{3+}$ (Sato and Saito, 1977)

The lel_2ob isomer, $(-)_{589}[Co(+chxn)_2(-chxn)]^{3+}$ (Fig. 4.10) has an approximate two-fold axis of rotation through the cobalt atom and the midpoint of the C–C bond in the ob chelate ring. The absolute configuration is $\Lambda(\delta\delta\lambda)$ or in full, Λ-$(-)_{589}[Co\{(S,S)(+)chxn\}_2 \{(R,R)(-)chxn\}\delta\delta\lambda]^{3+}$. Each chelate ring has an unsymmetrical skew conformation. The dihedral angles about the C–C bond in the chelate ring are $53°$ on the average. The two C–C bonds in the lel chelate rings are inclined at a mean angle of $3.9°$ with respect to the pseudo-three-fold axis, while that in the ob ring makes an angle of $64.4°$. The non-bonded short hydrogen-hydrogen contacts occur between NH_2 and CH groups in adjacent chelate rings. The average H ... H distance is 2.46 Å between the lel ring and the ob ring. The inclination angle of the co-ordination plane formed by Co and the two N atoms of the ob ring with respect to the pseudo-three-fold axis of the ion, $35.7°$, is significantly greater than those of the lel_3- and ob_3-isomers ($31.8°$ and $31.5°$, respectively). This difference in inclination angle may alleviate the non-bonded hydrogen-hydrogen interactions.

The $\Delta G°$ at 373 K, pH = 7.0 are as follows (Harnung, Sørensen, Creaser, Maegaard, Pfenninger and Schäffer, 1976):

$\Delta G°$ ($lel_2ob \rightarrow lel_3$)	-0.93 kJ mol^{-1}
$\Delta G°$ ($ob_2lel \rightarrow lel_3$)	-3.72
$\Delta G°$ ($ob_3 \rightarrow lel_3$)	-8.20

The complex ion, $(-)_{589}[Co(+chxn)(-chxn)_2]^{3+}$, has the absolute configuration $\Lambda(\delta\lambda\lambda)$ and is the $lelob_2$ isomer. It possesses a two-fold axis through the cobalt atom, bisecting the C–C bond in the lel chelate ring (Shintani, Sato and Saito, 1979).

trans-1,2-Diaminocyclopentane

The tris complex involving this ligand was recorded as early as 1928 (Jaeger and Blumendal, 1928). It was, however, suggested later that its existence was doubtful, since the chelate ring system involves much strain, when a molecule of *trans*-1,2-diaminocyclopentane is co-ordinated to a cobalt atom. In fact, the N ... N distance

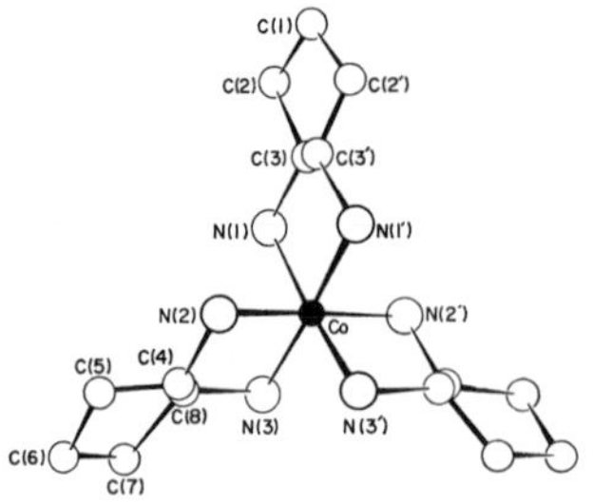

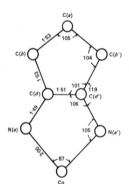

Fig. 4.11. $(-)_{589}[Co(+cptn)_3]^{3+}$ (Ito, Marumo and Saito, 1971)

of 3.14 Å in a free state of the ligand must decrease in length to 2.76 Å, on forming a five-membered chelate ring (Phillips and Royer, 1965). The existence of the tris-bidentate complex was established by the crystal structure analysis of $(-)_{589}[Co(+cptn)_3]$ Cl$_3$ · 4H$_2$O (Ito, Marumo and Saito, 1971). Fig. 4.11 shows a perspective drawing of the complex ion $(-)_{589}[Co(+cptn)_3]^{3+}$. The cobalt atom has a slightly distorted octahedral co-ordination of six nitrogen atoms with an average distance of 2.00 Å. The complex ion possesses an approximate D_3 symmetry. The conformation

Fig. 4.12. Bond lengths and angles of a chelate ring averaged by assuming D_3 symmetry

Table 4.3. Comparison of molecular geometry of the ligand cptn in the crystal with that calculated by energy minimization

	Observed in the [Co(cptn)$_3$]$^{3+}$	Calculated for free ligand
C(a)...C(b)	1.53 Å	1.53 Å
C(b)...C(d')	1.53	1.52
C(d)...C(d')	1.51	1.52
N(a)...C(d)	1.49	1.49
C(b)C(a)C(b')	105°	106°
C(a)C(b)C(d)	104	104
C(b)C(d)C(d')	101	102
N(a)C(d)C(d')	106	112
N(a)C(d)C(b)	119	112

of the cyclopentane ring is half-chair. The NCoN angles in the chelate ring average 86.7°. The distortion of the octahedron formed by the six nitrogen atoms is nearly the same as that for $[Co(en)_3]^{3+}$. The geometry of the chelate ring is illustrated in Fig. 4.12. The bond lengths are normal, however, the bond angles C(b′)C(d′)C(d) and N(a′)C(d′)C(b′), both deviate largely from the normal tetrahedral angle. Table 4.3 compares the results of strain energy minimization for a free ligand with the observed bond lengths and angles in the chelate ring. Large discrepancies in NCC angles are clearly due to chelation and hence a fused ring formation. This is the lel_3 isomer and the absolute configuration of the complex ion can be designated $\Lambda(\delta\delta\delta)$.

Sarcosine   CH$_2$—NHCH$_3$
|
CO$_2$H

The stereochemistry of sarcosinatobis(ethylenediamine)cobalt(III) complex ion was first studied as early as 1924 (Meisenheimer, Angermann and Holsten, 1924). They reported that four isomers could be isolated. When this work was repeated carefully only two forms were obtained (Buckingham, Mason, Sargeson and Turnbull, 1966). The second study suggested that sarcosinate ion was co-ordinated stereospecifically about one configuration of the Co(en)$_2$ moiety. In fact an X-ray study of $(-)_{589}$-[Co(sar)(en)$_2$]I$_2$ · 2H$_2$O (Blount, Freeman, Sargeson and Turnbull, 1967) revealed that the Co(sar) ring is slightly puckered, the conformation being λ.[11] The two Co(en) rings take δ and λ conformations respectively. These combinations presumably minimize the H(methyl)...H(amino) interactions. The absolute configuration of the entire complex ion can be fully designated as $\Delta(\lambda_{sar}\lambda_{en}\delta_{en})$. The absolute configuration of the co-ordinated secondary nitrogen atom is S. In this stable Δ-S form the hydrogen atom is balanced over the adjacent Co(en) ring and the methyl group lies in the space between the two cobalt-ethylenediamine rings. Recently Fujita, Yoshikawa and Yamatera (1976) succeeded in separating all the four possible isomers by the chromatographic method with an SP-Sephadex column. The crystal structure of the less stable isomer has not yet been determined. From the formation ratio the Δ-[Co(R-sar)(en)$_2$]$^{2+}$ isomer is only 3.8 kJ mol^{-1} less stable than the Δ-[Co-(S-sar)(en)$_2$]$^{2+}$ isomer. This experimental value of the ΔG difference is about half of the calculated value of 7.1 kJ mol^{-1} on strain energy minimization (Anderson, Buckingham, Gainsford, Robertson and Sargeson, 1975).

***

11 In $(+)_{436}$[Co(sar)(NH$_3$)$_4$]$^{2+}$ the Co(sar) ring is puckered, too (Larsen, Watson, Sargeson and Turnbull, 1968).

N-methyl-(S)-alanine CH_3
$$\underset{\underset{\text{CO}_2\text{H}}{|}}{\overset{\overset{\text{CH}_3}{|}}{\text{H–C–NHCH}_3}}$$

In Δ-R-[Co(N-Me-(S)-ala)(en)$_2$]$^{2+}$ (Anderson, Buckingham, Gainsford, Robertson and Sargeson, 1975), the two cobalt-ethylenediamine rings take δ and λ conformations. The amino acid contains *trans* methyl groups with the R and S configurations about the N and C centres, respectively. The Co–N–C(methyl) angle of 120.2° differs markedly from the regular tetrahedral angle due to the steric hindrance of the adjacent Co-en ring. This observed geometry of the complex ion agrees satisfactorily with the result of strain energy minimization. The calculated strain energy increases in the order: N(S)–C(R) < N(R)–C(S) < N(S)–C(S) < N(R)–C(R) for Δ isomers. This relationship results in large part from the non-bonded interactions between the ethylenediamine ring and/or the methyl group and the carbon atom of the amino acid ligand. In fact, in Δ-[Co(N-Me–(S)-ala)(en)$_2$]$^{2+}$ mutarotation about the C centre of the alaninato chelate occurs at pH > 12 and at equilibrium the Δ-R : Δ-S ratio was ca. 4.

C. Six-Membered Chelate Rings

The conformational problem presented by complexes involving six-membered chelate rings is similar to that posed by cyxlohexane, except that the ligand-metal-ligand angle is nearly 90°. Three possible conformations exist for a single metal-trimethylene-diamine ring, one rigid chair form with mirror symmetry and two enantiomeric twist-boat forms with a two-fold axis of rotation. The third conformation, the boat form with mirror symmetry, cannot be accommodated to form a tris-bidentate complex and is rarely observed in the structure of metal-chelate complexes. For each of the twist-boat conformations of the Co-tn ring two skew lines can be defined: one through the two nitrogen atoms and the other through the carbon atoms that are bonded to the two nitrogen atoms. According to the helicity associated with these lines the two enantiomeric forms are labelled δ and λ.

Niketić and Woldbye (1973a) showed that 16 possible conformers exist of the [M(tn)$_3$] system for each of the absolute configurations Δ and Λ, in which the chelate rings adopt any of the three stable conformations mentioned above. Three of these, chair$_3$, twist-boat$_3$ with *lel*$_3$, and twist-boat$_3$ with *ob*$_3$ conformations are called homoconformational and the remaining 13 forms are termed hetero-conformational.

When a M-tn ring takes a chair conformation in a tris-bidentate complex, the ring in question may adopt one of the two orientations which give rise to two conformers if, and only if, the two other rings are *not* related by a two-fold axis of rotation (Fig. 4.13).

Raymond designated these two ring conformations as p and a, depending on whether the fold direction of the ring, determined by the orientation of the C–C–C plane, is parallel or antiparallel to the chirality defined by the arrangement of the

Fig. 4.13. Two tris-bidentate complexes with Λ absolute configuration are shown. When the two rings A and B are not related by a two-fold axis of rotation, the folding direction of the third ring with chair conformation can be distinguished as p or a

three chelate rings (left-handed for Λ absolute configuration (Jurnak and Raymond, 1972).

The absolute configuration of $(+)_{589}[Co(tn)_3]^{3+}$ was determined with the two isostructural bromide and chloride monohydrates (Nomura, Marumo and Saito, 1969; Nagao, Marumo and Saito, 1973). Figure 4.14 presents the absolute configuration of $(+)_{589}[Co(tn)_3]^{3+}$. It has an approximate three-fold axis and the three chelate rings take chair conformation. The three six-membered chelate rings are nearly but not exactly identical, reflecting the flexibility of the six-membered chelate ring. The chelate ring is rather flattened out due to non-bonded hydrogen interactions. Bond angles CoNC are much larger than the normal tetrahedral angle, the average being $122.0°$. The mean NCoN angle in the chelate ring is $91.0°$. Ellipsoids of thermal motion of one of the chelate rings which is most loosely packed in the crystal indicated that the largest amplitude of thermal vibration of the carbon atoms is primarily perpendicular to the plane formed by the two bonds (C–C or C–N) for each atom. This large thermal motion suggests a conformational equilibrium involving significant amounts of two or more conformers in solution at room temperature (Beddoe, Harding, Mason and Peart, 1971). Unlike the homoconformational complex ion, $[Co(tn)_3]^{3+}$, the complex ion, $[Cr(tn)_3]^{3+}$ has a two-fold axis of rotation in $[Cr(tn)_3][Ni(CN)_5] \cdot 2H_2O$. One chelate ring takes a twist-boat conformation and the remaining two are chair forms, i.e. the complex ion can be designated $\Lambda(ap\delta)$ (syn-chair$_2$*lel* conformation) and its enantiomer (Jurnak and Raymond, 1974). The absolute configuration of $(-)_{589}[Co(acac)(tn)_2]^{2+}$ is Δ and the two Co-tn rings take a chair conformation. This is an anti-chair$_2$ isomer, $\Delta(pp)[\equiv \Delta(aa)]$ (Matsumoto, Kawaguchi, Kuroya and Kawaguchi, 1973).

Fig. 4.14. $(+)_{589}[Co(tn)_3]^{3+}$ (Nagao, Marumo and Saito, 1973)

2,4-Diaminopentane

2,4-Diaminopentane acts as a bidentate to form a six-membered chelate ring. There are three isomers since the ligand has two asymmetric carbon atoms:

$$
\begin{array}{ccc}
& CH_3 & \\
& H_2N-C-H & \\
R,R & CH_2 & \\
& H-C-NH_2 & \\
& CH_3 &
\end{array}
\qquad
\begin{array}{ccc}
& CH_3 & \\
& H-C-NH_2 & \\
S,S & CH_2 & \\
& H_2N-C-H & \\
& CH_3 &
\end{array}
\qquad
\begin{array}{ccc}
& CH_3 & \\
& H-C-NH_2 & \\
R,S(meso) & CH_2 & \\
& H-C-NH_2 & \\
& CH_3 &
\end{array}
$$

Table 4.4 summarises the possible conformations of the chelate rings formed by 2,4-ptn. The structures of three homoconformational isomers have been determined. The stable conformation of the chelate ring formed by *meso*-ptn may be a chair form with the two methyl groups in equatorial positions (See Table 4.8) (Appleton and Hall, 1971). With three chairs Λ-[Co(*meso*-ptn)$_3$]$^{3+}$ may adopt two conformations: Λ-(ppp)[$\equiv \Lambda$(aaa)] and Λ-(paa)[$\equiv \Lambda$(app)]. The former has a three-fold axis of symmetry and the latter lacks any symmetry element. The two isomers were synthesized and separated by Kojima and Fujita (1976). These authors designated the C_3 and C_1 isomers as *fac* and *mer* respectively. Both isomers were resolved into optical

Table 4.4. Disposition of the two C–CH$_3$ bonds in the chelate rings formed by the isomers of 2,4-diaminopentane

Ligand	Conformation		
	Chair	δ-twist-boat	λ-twist-boat
R,R-2,4-ptn	a, e	a, a	e, e
S,S-2,4-ptn	a, e	e, e	a, a
R,S-2,4-ptn	a, a or e, e	a, e	a, e

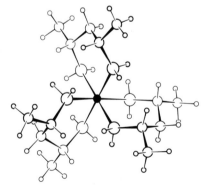

Fig. 4.15. $(+)_{589}$-*fac*-[Co(meso-ptn)$_3$]$^{3+}$ (Sato and Saito, 1978)

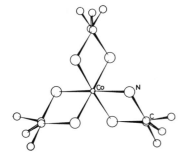

Fig. 4.16. $(-)_{546}[Co(R,R\text{-ptn})_3]^{3+}$ (Kobayashi, Marumo and Saito, 1973)

isomers. Figure 4.15 presents the complex ion, $(+)_{589}$-*fac*-[Co(*meso*-ptn)$_3$]$^{3+}$ (Sato and Saito, 1978). The three chelate rings take the stable chair conformation with the C–CH$_3$ bond equatorial as expected. Unlike [Co(tn)$_3$]$^{3+}$, no anomalous thermal vibration of the ring carbon atoms was observed owing to substituted methyl groups. The geometry of the Co(tn)$_3$ portion is similar to that in [Co(tn)$_3$]$^{3+}$. An interesting feature is that the geometries of the three chelate rings are slightly but significantly different from each other. This may be due partly to the specific packing forces in the crystal lattice and partly to the flexibility of the six-membered chelate rings. As the NCoN angle in the chelate ring increases from 87.8° to 92.7°, the CCC angle decreases from 117.0° to 113.8°. Corresponding to this change in opposite bond angles in the chelate ring, the inner fragment (C–N–Co–N–C) flattens with slight puckering of the outer portion (N–C–C–C–N), namely the dihedral angle between the NCoN and NCCN planes increases from 144.5° to 156.8° and that between the NCCN and CCC planes decreases from 125.1° to 117.8°.

Structures of $(-)_{546}[Co(R,R\text{-ptn})_3]^{3+}$ and $(+)_{546}[Co(R,R\text{-ptn})_3]^{3+}$ are shown in Figs. 4.16 and 4.17 respectively. In both cases, the chelate ring takes a λ-skew-boat conformation with the two methyl groups in equatorial positions as shown in Table 4.4. The former is Δ(λλλ), the *lel*$_3$ isomer (Kobayashi, Marumo and Saito, 1973), while the latter Λ(λλλ), the *ob*$_3$ isomer (Kobayashi, Marumo and Saito, 1972b). Conformational analysis of tris-diamine complexes has been carried out by several authors, notably by Woldbye and his coworkers (Geue and Snow, 1971; Niketić and Woldbye, 1973a, b, 1974; Niketić, 1974). The calculation reproduced the molecular geometry fairly satisfactorily; however, the calculated strain energies for different isomers by different authors were sometimes inconsistent. Recently, a fast convergent minimization programme was applied to calculate equilibrium conformations of all possible conformers of [Co(tn)$_3$]$^{3+}$ and ten species of [Co(ptn)$_3$]$^{3+}$

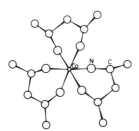

Fig. 4.17. $(+)_{546}[Co(R,R\text{-ptn})_3]^{3+}$ (Kobayashi, Marumo and Saito, 1972b)

Table 4.5. Observed and calculated bond lengths and angles in $[Co(tn)_3]^{3+}$ with C_3-chair$_3$ conformation

	Observed in $[Co(tn)_3]^{3+a}$	Calculated[b]
Co–N	1.979 ± 0.003 Å	2.045 ± 0.01 Å
N–C	1.484 ± 0.006	1.477 ± 0.03
C–C	1.499 ± 0.006	1.543 ± 0.01
NCoN	90.4 ± 0.1°	90.0 ± 4.0°
CoNC	122.2 ± 0.2	120.0 ± 2.0
NCC	111.8 ± 0.3[c]	111.5 ± 0.7
CCC	113.5 ± 0.3	110.8 ± 1.0

a   Nagao, Marumo and Saito, 1973; averaged assuming C_3 symmetry.
b   Niketić, Rasmussen, Woldbye and Lifson, 1976. + and − indicate the upper and the lower limits respectively of the values found in all minimum-energy conformations based on the different force fields adopted.
c   One anomalous angle due to anisotropic thermal vibration has been omitted from averaging.

Table 4.6. Average bond lengths and angles in homoconformational $[Co(ptn)_3]^{3+}$ ions

	fac-chair$_3$		lel$_3$		ob$_3$	
	obs[a]	calc[d]	obs[b]	calc[d]	obs[c]	calc[d]
Co–N	1.999	2.043	1.985	2.043	1.988	2.054 Å
N–C	1.498	1.477	1.489	1.479	1.50	1.478
C–C	1.510	1.544	1.516	1.549	1.53	1.550
C–C$_{Me}$	1.528	1.549	1.530	1.548	1.50	1.549
NCoN	92.7	93.85	89.1	88.10	87.9	87.80°
CoNC	122.9	120.13	118.0	114.10	115.7	114.62
NCC	110.8	110.68	112.0	111.83	109.0	111.96
CCC	113.8	110.21	117.3	112.64	116.7	113.24
NCC$_{Me}$	109.5	109.38	109.4	110.26	112.2	110.24
CCC$_{Me}$	112.7	109.53	111.0	109.38	105.7	109.26

a   Sato and Saito, 1978, averaged for one chelate ring with an obtuse NCoN angle assuming mirror symmetry.
b   Kobayashi, Marumo and Saito, 1973; averaged assuming D_3 symmetry.
c   Kobayashi, Marumo and Saito, 1972b; averaged assuming D_3 symmetry.
d   Niketić, Rasmussen, Woldbye and Lifson, 1976.

(Niketić, Rasmussen, Woldbye and Lifson, 1976). The calculations demonstrated that various previously suggested conformations of the same type converge to a common equilibrium conformation having the highest possible symmetry. The C_3-chair$_3$ conformer represents the gobal minimum in the selected force field. Table 4.5 compares the observed geometry of $(+)_{589}[Co(tn)_3]^{3+}$ with the result of strain energy minimization. The agreement in Table 4.5 indicates that the geometry of the complex is well reproduced.

Table 4.6 compares the mean bond lengths and angles in homoconformational isomers of the $[Co(ptn)_3]^{3+}$ ions with the results of strain energy minimization. As seen from Table 4.6 the geometries of the complex ions are fairly well reproduced. Various energy contributions of strain energies are listed in Tables 4.7 and 4.8. Detailed analysis of the 16 isomers revealed that no correlation exists between the various contributions from bond length and angle distortions, torsional strain and non-bonded interactions. The energy minima are reached by very delicate balancing of all energy contributions through adjustment of practically all internal variables in these flexible complex ions. The longer non-bonded interactions contributed markedly to the stabilization. For example, in the case of the minimum global conformation of $M(tn)_3$, the non-bonded interactions shorter than 3 Å contributed 42.2 kJ mol^{-1} and the longer ones -49.7 kJ mol^{-1}. In the case of homoconformational isomers listed in Tables 4.7 and 4.8, the chair$_3$ conformations have relatively high angle deformation energies, owing to the flattening of the chelate rings and low torsional energies and torsional strain. The relative strain energies are in the order:

$$C_3\text{-chair}_3 < lel_3 < ob_3.$$

Table 4.7. Energy contributions for homoconformational $[Co(tn)_3]^{3+}$

	fac-chair$_3$	lel$_3$	ob$_3$
Bond stretching deformations	6.7[a]	6.5	9.7
Bond angle deformations	39.3	18.8	25.6
Torsional strain	11.7	29.9	31.8
Non-bonded interactions	−7.5	−1.7	9.3
Total conformational energy	50.2	53.5	76.4
Difference	0.0	3.3	26.2

a   All energies are given in kJ mol^{-1}.

Table 4.8. Energy contributions for homoconformational $[Co(ptn)_3]^{3+}$ isomers with the two methyl groups in equatorial positions

	fac-chair$_3$	lel$_3$	ob$_3$
Bond stretching deformations	7.6[a]	7.8	11.5
Bond angle deformations	39.6	19.4	25.2
Torsional strain	12.6	29.5	31.4
Non-bonded interactions	−19.1	−14.6	−3.8
Total conformational energy	40.7	42.1	64.3
Difference	0.0	1.4	23.6

a   All energies are given in kJ mol^{-1}.

Malonate ion

The chelated malonato-metal ring has a high degree of conformational flexibility. In the structure of $\Delta[\text{Co}(R-\text{pn})_3]\Lambda[\text{Cr}(\text{mal})_3] \cdot 3H_2O$, the three malonato-Cr(III) rings are equivalent by symmetry and possess an envelope conformation in which only the methylene carbon atom is significantly displaced from the plane of the chelate ring (Butler and Snow, 1971,1976). In the diastereoisomer $(+)_{546}[\text{Co}(\text{mal})_2(\text{en})](-)_{589}$-$[\text{Co}(\text{NO}_2)_2(\text{en})_2]$ both rings are reported as having an approximately planar conformation. The largest deviation from the mean plane of the Co-mal ring is 0.23 Å of a ligating oxygen atom and the pattern for both rings suggests a distortion towards a skew-boat conformation (Matsumoto and Kuroya, 1971, 1972).

D. Seven-Membered Chelate Rings

Only a few structures containing seven-membered chelate rings are known. One important member of the bidentates forming a seven-membered chelate ring is 1,4-diaminobutane(tetramethylenediamine). The crystal structure of $(+)_{589}[\text{Co}(\text{tmd})_3]$-$\text{Br}_3$ has been determined (Sato and Saito, 1975). Figure 4.18 shows a perspective drawing of the complex ion. It has an approximate D_3 symmetry. In Fig. 4.19 chelate ring is puckered and chiral. The ring conformation can be designated as λ, providing the helicity is defined by the line joining the two ligating nitrogen atoms and the line joining the two carbon atoms bonded to the nitrogen atoms. The absolute configuration is $\Delta(\lambda\lambda\lambda)$, lel_3. The Co–N–C–C and N–C–C–C groups take δ and λ conformations respectively. The seven-membered chelate ring is strained: the mean CoNC, NCC and CCC angles are 122.9, 113.0 and 116.1° respectively. The NCoN angle in the chelate ring is, however, close to 90°, the mean value being 89.2°.

The ornithinate ion, $H_2N(CH_2)_3 \cdot CH(NH_2)CO_2^-$, acts as a bidentate on chelation to palladium(II) to form seven-membered chelate rings. In crystals of bis(S-

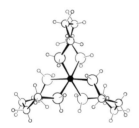

Fig. 4.18. $(+)_{589}[\text{Co}(\text{tmd})_3]^{3+}$ (Sato and Saito, 1975)

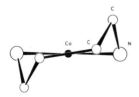

Fig. 4.19. A projection of the chelate ring along the two-fold axis (Sato and Saito, 1975)

ornithinato)palladium(II), the chelate rings take a twist-chair conformation with the carboxylate group in a quasi-equatorial position (Nakayama, Matsumoto, Ooi and Kuroya, 1973).

2 Terdentates

Structures involving three types of terdentates are known: i) linear, ii) branched and iii) cyclic.

Diethylenetriamine

This linear terdentate is a strong chelating agent and three isomers are possible as illustrated in Fig. 4.20. Among the three geometric isomers, the *u-facial-* and *mer-* isomers are optically active and have pairs of catoptromers respectively, whereas the *s-facial* isomer is optically inactive. All the geometric and optical isomers in this system have been isolated and characterised (Keene, Searle, Yoshikawa, Imai and Yamasaki, 1970). Structures of *s-facial* and *u-facial*-[Co(dien)$_2$]$^{3+}$ are known. In the complex ion, *s-fac*-[Co(dien)$_2$]$^{3+}$, the ligand co-ordinates to the cobalt atom with the terminal amine group in *cis*-positions. The five-membered chelate rings take an unsymmetrical skew conformation (Kobayashi, Marumo and Saito, 1972c). The complex ion has a centre of symmetry and has an approximate mirror plane through the cobalt and the two secondary nitrogen atoms and bisecting the NCoN angle formed by the two terminal nitrogen atoms of a ligand. Figure 4.21 illustrates the two con-

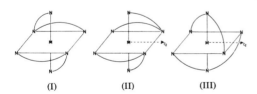

(I)   (II)   (III)

Fig. 4.20. Schematic drawings of (I) *s-facial-* (II) *u-facial-* and (III) *mer-*isomer of [Co(dien)$_2$]$^{3+}$

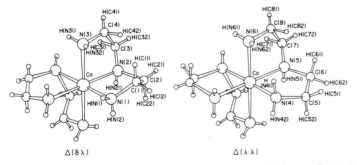

Δ(δλ)   Δ(λλ)

Fig. 4.21. Perspective drawings of the two conformers of $(-)_{589}$-*u-fac*-[Co(dien)$_2$]$^{3+}$ (Konno, Marumo and Saito, 1973)

formers observed in $(-)_{589}$-u-fac-[Co(dien)$_2$][Co(CN)$_6$] · 2H$_2$O (Konno, Marumo and Saito, 1973). The complex ions have a two-fold axis of rotation. The absolute configuration can be designated as skew chelate pairs ΔΛΔ. However, the conformations of the two chelate rings formed by a dien molecule in one complex ion are δλ, while those in the other are λλ. The two fused chelate rings with conformations δλ have eclipsed envelope and symmetrical skew conformations, whereas those with λλ conformations are both eclipsed envelope conformations.

Yoshikawa (1976) carried out the conformational analysis of the [Co(dien)$_2$]$^{3+}$ system based on Boyd's procedure (1968). Table 4.9 shows the observed and calcu-

Table 4.9. Average bond lengths and angles in [Co(dien)$_2$]$^{3+}$ isomers

	s-fac		u-fac-1[e]		u-fac-2[e]		mer
	obs[b]	calc[d]	obs[c]	calc	obs[c]	calc	calc
Co−N	1.97	1.966	1.957	1.958	1.951	1.952	1.976 Å
			1.969	1.968	1.968	1.973	
Co−N[a]	1.95	1.944	1.970	1.960	1.970	1.971	1.942
C−N	1.48	1.491	1.491	1.490	1.488	1.488	1.495
C−N[a]	1.51	1.494	1.496	1.495	1.499	1.497	1.486
C−C	1.51	1.515	1.516	1.513	1.509	1.514	1.514
NCoN	86.6	88.3	85.4	87.0	84.9	86.8	86.0°
	87.2	88.9	86.4	88.3	85.7	87.0	86.1
CN[a]C	116.0	113.8	113.4	113.0	110.4	110.3	114.5

a   Secondary nitrogen atom.
b   Kobayashi, Marumo and Saito, 1972c.
c   Konno, Marumo and Saito, 1973.
d   Yoshikawa, 1976.
e   u-fac-1: ΔΛΔ(δλ) and its catoptromer.
    u-fac-2: ΔΛΔ(λλ) and its catoptromer.

Table 4.10. Energy contributions for the [Co(dien)$_2$]$^{3+}$ system

	s-fac	u-fac-1[a]	u-fac-2[a]	mer
Bond stretching deformations	3.3	3.5	4.6	4.6
Bond angle deformations	9.6	10.0	11.7	22.8
Torsional strain	32.9	33.1	29.8	22.6
Non-bonded interactions	25.4	25.6	26.1	21.4
Total conformational energy	71.2	72.2	72.2	71.4
Difference	0.0	1.0	1.0	0.2

All energies are in kJ mol^{-1}.
a   u-fac-1: ΔΛΔ(δλ) and its catoptromer.
    u-fac-2: ΔΛΔ(λλ) and its catoptromer.

lated interatomic distances and bond angles. The agreement in the Table is good, because the energy parameters have been adjusted for better fitting of the minimized structures with the observed ones. The final energy contributions are tabulated in Table 4.10. The total strain energies of the isomers differ very little. It is not possible that the reliability of the strain energy differences could be better than ± 8 kJ mol^{-1}. In fact, the formation ratios experimentally determined in the equilibrium mixture of bromides at 298 K were $s\text{-}fac : u\text{-}fac : mer = 7:30:63$. The discrepancy may be ascribed to an inadequate choice of energy parameters, neglect of longer non-bonded interactions and/or zero level imbalances (Dwyer and Searle, 1972; Dwyer, Geue and Snow, 1973). However, the energy parameters were successfully applied for energy minimization of the $[Co(linpen)]^{3+}$ system (see p. 83).

1,1,1-Tris(aminoethyl)ethane

$$CH_3-C\begin{array}{l} \diagup CH_2-NH_2 \\ -CH_2-NH_2 \\ \diagdown CH_2-NH_2 \end{array}$$

The bis complex of Co(III) has its non-ligating atoms above and below the trigonal planes of the ligating nitrogen atoms, whereas the tris(bidentate) complexes, such as $[Co(en)_3]^{3+}$, have the non-ligating atoms between opposed trigonal planes. The IUPAC scheme for the designation of the absolute configuration cannot be directly applied for assigning a label to this type of complex. The crystal structure of $(+)_{589}$-$[Co(tame)_2](+)_{589}[R,R\text{-tart}] \cdot xH_2O$ has been determined (Geue and Snow, 1977). The complex cation is shown in Fig. 4.22. It has an approximate D_3 symmetry. The conformation of all the six six-membered chelate rings is λ and is intermediate between that of the regular skew-boat and a boat form and may be described as an asymmetric skew-boat. When this complex is viewed down its triad axis, the atoms of the ligands form a right-handed helix, hence this complex ion can be fully designated as $\Delta\lambda\lambda\text{-}[Co(tame)_2]^{3+}$. Three conformational isomers are possible, which can be designated $\lambda\lambda$, $\delta\delta$ and $\delta\lambda$ or $\lambda\delta$. The minimum strain energy form of the $\delta\delta$ (or $\lambda\lambda$) isomer has D_3 symmetry and the $\lambda\delta$ (or $\delta\lambda$) form, C_{3i}. These isomers may interconvert by a trigonal twist of one ligand, whereby the CH$_2$ groups move from

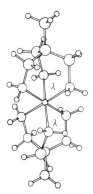

Fig. 4.22. $\Delta\lambda\lambda\text{-}[Co(tame)_2]^{3+}$

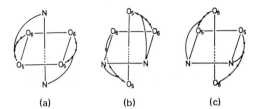

Fig. 4.23. Schematic drawings of **(a)** the trans(N), **(b)** the *cis*(N)-*trans*(O_5) and **(c)** the *cis*(N)-*trans*(O_6) isomers of the [Co(*S*-asp)_2]⁻ ion (Oonishi, Shibata, Marumo and Saito, 1973)

(a)   (b)   (c)

one side of the Co-N-quarternary C plane to the other. The strain energy minimization indicates that the racemic isomer is more stable by 6.7 kJ mol^{-1}. It is anticipated that the interconversion may be rapid in solution.

S-Aspartic acid

When *S*-aspartic acid, $HO_2C–CH(NH_2)–CH_2–CO_2H$, acts as a terdentate to form an octahedral complex, there are three possible isomers as shown in Fig. 4.23. O_5 and O_6 refer to those oxygen atoms that form five- and six-membered chelate rings with amino N atoms respectively. Figure 4.24 shows *cis*(N)-*trans*(O_5) and *cis*(N)-*trans*(O_6) isomers (Oonishi, Shibata, Marumo and Saito, 1973; Oonishi, Sato and Saito, 1975). Two aspartic acid residues are octahedrally co-ordinated to a cobalt atom through two amino N atoms and four carboxylic O atoms. Five-membered, six-membered and seven-membered chelate rings are formed. Two N atoms are in *cis*-positions. The co-ordination octahedrons are slightly distorted, probably owing to the non-bonded hydrogen interactions as well as to the formation of strained chelate rings.

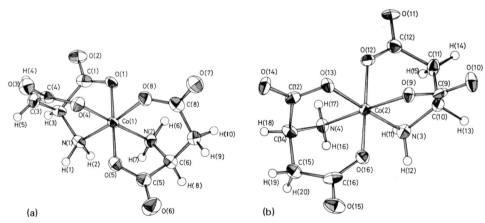

(a)   (b)

Fig. 4.24. (a) *cis*(N)-*trans*(O_5)- and (b) *cis*(N)-*trans*(O_6)-[Co(*S*-asp)_2]⁻ (Oonishi, Shibata, Marumo and Saito, 1973)

S-2,3-Diaminopropionic acid $HO_2C-CH(NH_2)-CH_2-NH_2$

The ligand can form a bis complex ion with Co(III) analogous to $[Co(S\text{-asp})_2]^-$.
There are three possible isomers: i) *trans, cis, cis* ii) *cis, cis, trans* and iii) *cis, trans, cis*, where the carboxyl groups are designated first, then the α-amino group and finally the β-amino group. They all have catoptromers. When *R*- and *S*-2,3-diaminopropionate ions co-ordinate, two more geometric isomers are possible (Freeman and Liu, 1968). The structure and absolute configuration of $(-)_{546}[\overset{\centerdot}{Co}(C_3H_7N_2O_2)_2]Br$ has been determined (Liu and Ibers, 1969). The crystal was proved to contain an *S-cis, trans, cis*-isomer. X-ray determination was necessary since the assignment of the absolute configuration was impossible on the basis of circular dichroism spectra.

Sarcosinate-N-monopropionic Acid

The O,N,O-terdentate ligand, $HO_2C-CH_2-N(CH_3)-CH_2-CH_2-CO_2H$, co-ordinates facialy to the metal atom. In $(+)_{546}$-*cis*(O)-$[Co(sarmp)(NH_3)_3]^+$, the resultant six-membered chelate ring takes the skewboat conformation, the $N-CH_3$ and C=O bonds both being in equatorial positions. The five-membered chelate ring assumes an asymmetric envelope form with λ conformation. The absolute configuration of the N atom is *R* (Okamoto, Tsukihara, Hidaka and Shimura, 1973).

Tribenzo[b,f,j]-[1,5,9]triazacyclododecahexaene

The ligand is the trimer formed by the self-condensation of o-aminobenzaldehyde in the presence of metal ions. It is stereochemically rigid. The complex ion, $[Co(TRI)_2]^{3+}$ has quite a different geometric array of chelate rings compared to the familiar

$[Co(en)_3]^{3+}$ ion. The absolute configuration of $(+)_{546}[Co(TRI)_2]^{3+}$ has been determined (Wing and Eiss, 1970). This is a six-co-ordinate octahedral complex. The cobalt atom is sandwiched midway between the two parallel planes (2.36 Å apart) formed by three N atoms. The three N atoms of each ligand form an equilateral triangle; the upper one is rotated counterclockwise by 8° with respect to the lower triangle from the position expected for a regular octahedron, as in the case for Λ-$(+)_{589}[Co(en)_3]^{3+}$. The TRI ligand is propeller shaped, the mean pitch of the planar benzene being 14° with respect to the plane formed by the N atoms. The benzene pairs are opened away from each other at the perimeter of the complex with an average angle between benzene pairs of 15°.

$R(-)$-2-Methyl-1,4,7-triazacyclononane

The cyclic ligand also acts as a terdentate and gives an analogous bis-complex with cobalt(III). The crystal structure of $(-)_{589}[Co(R\text{-MeTACN})_2]I_3 \cdot 5H_2O$ has been determined (Mikami, Kuroda, Konno and Saito, 1977). The complex ion is disordered on a site of D_3 symmetry around the triad axis. Two molecules of the cyclic

$$
\begin{array}{c}
\text{NH} \\
\diagup \qquad \diagdown \\
(\text{CH}_2)_2 \qquad (\text{CH}_2)_2 \\
\diagup \qquad\qquad \diagdown \\
\text{HN–CH–CH}_2\text{–NH} \\
| \\
\text{CH}_3
\end{array}
$$

ligand co-ordinate to the Co atom with six secondary N atoms from above and below the Co atom to form an octahedral complex. There are six five-membered chelate rings with λ conformation. The C–CH_3 bond is equatorial with respect to the mean plane of the chelate ring. The $[CoN_6]$ chromophore is elongated and twisted around the triad axis. The direction of twist is the same as that of $\Delta(-)_{589}$-$[Co(en)_3]^{3+}$.

3 Quadridentates

Structures containing linear, branched and cyclic quadridentates have been extensively studied. The linear quadridentates are largely aliphatic tetramines:

$$H_2N-(CH_2)_m-NH-(CH_2)_n-NH-(CH_2)_p-NH_2 \qquad m,n,p = 2,3$$

When $m = n = p = 2$, the ligand is triethylenetetramine (trien). Other related tetramine ligands are often abbreviated as m, n, p-tet and a prime indicates substituted methyl groups (cf. abbreviation of ligands on p. 10). A great number of complexes containing synthetic macrocyclic quadridentates has been prepared and their stereochemistry has been extensively studied notably by Busch and Curtis. The most fundamental macrocyclic ligand is 1,4,8,11-tetraazacyclotetradecane(cyclam):

$$
\begin{array}{c}
\text{CH}_2 \\
\diagup \quad \diagdown \\
\text{H}_2\text{C} \quad\ \text{CH}_2 \\
| \qquad\quad | \\
\text{HN} \quad\ \text{NH} \\
\diagup \qquad\quad \diagdown \\
\text{H}_2\text{C} \qquad\qquad \text{CH}_2 \\
| \qquad\qquad\qquad | \\
\text{H}_2\text{C} \qquad\qquad \text{CH}_2 \\
\diagdown \qquad\quad \diagup \\
\text{HN} \quad\ \text{NH} \\
| \qquad\quad | \\
\text{H}_2\text{C} \quad\ \text{CH}_2 \\
\diagdown \quad \diagup \\
\text{CH}_2
\end{array}
$$

These complexes will not be described here, since a number of excellent review articles on this subject have already been published (Busch, 1967; Curtis, 1968; Busch, Farmery, Goedken, Katovic, Melnyk, Sperari and Tokel, 1971; Lindoy and Busch, 1971).

1,8-Diamino-3,6-diazaoctane(triethylenetetramine)

The linear quadridentate can co-ordinate to the Co atom to form an octahedral complex. There are three possible geometrical isomers as shown in Fig. 4.25. The *cis-α* and *cis-β* isomers are chiral. Furthermore, there exist two conformers for the *cis-β* isomers arising from alternative configurations of the two secondary N atoms as illustrated in Fig. 4.26, where the conformations of the two chelate rings on the equatorial plane are different. In solution, Δ-*cis*-β-(R,R)- and Δ-*cis*-β-(R,S)-[Co(trien)-$(H_2O)_2]^{3+}$ ions mutarotated to the thermodynamically more stable conformer ($\sim$12.6 kJ mol^{-1}) and the Δ-*cis*-β(R,S) isomer could not be obtained in a stable crystalline form (Buckingham, Marzilli and Sargeson, 1967). The crystals of racemic *cis*-β-[CoCl(trien)(H$_2$O)] ClO$_4$ are composed of Λ-*cis*-β-(S,S) and Δ-*cis*-β-(R,R) isomers (Freeman and Maxwell, 1969). The two outer chelate rings have unsymmetrical skew conformation and the central one is envelope type. In contrast to this, a substituted trien ligand, 3(S)8(S)-2′,2,2′-tet, gives the *cis*-β-(R,S) isomer in a stable crystalline form, together with the *cis*-α-isomer (Yoshikawa, Saburi, Sawai and Goto, 1969). This is due to the equatorial preference of C–CH$_3$ bonds. Figure 4.27a–c illustrate the structure of the *cis*-β-, *cis*-α- and *trans*-[Co(NO$_2$)$_2$(3(S)8(S)-2′,2,2′tet)]$^+$ as revealed by X-ray analysis of

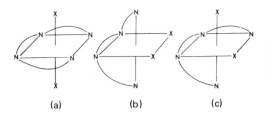

(a)   (b)   (c)

Fig. 4.25. Three possible cobalt-trien co-ordinations
(a) *trans*,   **(b)** *cis-α*,   **(c)** *cis-β*
      (*mm*)   (*ff*)   (*mf*)[12)]

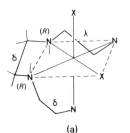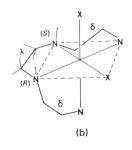

(a)   (b)

Fig. 4.26. (a) Δ-*cis*-β-(R,R) and (b) Δ-*cis*-β-(R,S) isomers

---

12 For the notation in parentheses see p. 83.

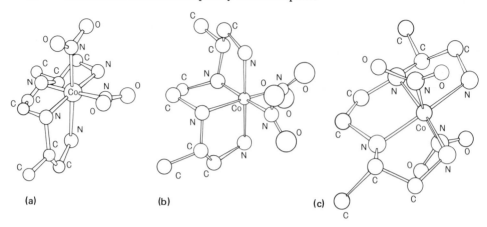

Fig. 4.27. (a) *cis-β-*, (b) *cis-α-* and (c) *trans-*isomers of [Co(NO$_2$)$_2$(3(*S*)8(*S*)-2',2,2'-tet)]$^+$ (Ito, Marumo and Saito, 1970; 1972a,b)

their perchlorate crystals (Ito, Marumo and Saito, 1970, 1972a, 1972b). In all these isomers, the C–CH$_3$ bonds are in equatorial positions. Accordingly the conformations of the five-membered chelate rings with substituted CH$_3$ group are δ and that of the central chelate ring is λ. The *cis-β* isomer takes the absolute configuration Δ and the absolute configurations of the two asymmetric N atoms are *R* and *S*. The apical chelate ring is in the eclipsed envelope conformation, while the in-plane terminal chelate ring has the two C atoms on the same side of the co-ordination plane. The central chelate ring has an unsymmetrical skew conformation. The steric strain which arises from the *cis-β* co-ordination of the ligand is partly alleviated by distortion of the chelate rings. The *cis-α* isomer has an approximate two-fold axis. The absolute configuration is Λ and the two N atoms have the absolute configuration *S*. The two outer chelate rings take the unsymmetrical skew conformation. The central chelate ring assumes nearly symmetrical skew conformation. The *trans-*isomer also possesses an approximate two-fold axis through the cobalt and bisecting the C–C bond in the central chelate ring. The ligand forms a girdle around the Co atom. The absolute configuration of the two secondary N atoms are *R*. The NCoN angles of the outer chelate rings are 85° and that of the central one is 88°. The ligand angular strain is further evidenced at the two asymmetric N atoms and at the two C atoms in the central chelate rings. The bond angles involving these atoms deviate largely from a regular tetrahedral angle. The two outer chelate rings have an unsymmetrical envelope conformation, whereas in the central ring the two C atoms are on the same side of the co-ordination plane. The conformational analysis of these complexes was carried out by Boyd's method (Ito, Marumo and Saito, 1972b). The bond lengths and angles in the complexes were reproduced within twice the standard deviation of the values obtained by structure analysis. The major angular distortions observed were accurately predicted from the minimization calculations. The final energy terms are listed in Table 4.11. The result indicates that the *cis-β* form is the most stable of the three isomers. This is supported by the observation that the *trans-*isomer is easily isomerised to the *cis-β* form by recrystallisation from water.

Table 4.11. Distribution of conformational strain energy in kJ mol^{-1}

	cis-α	cis-β	trans
Bond length deformations	2.5	5.4	2.9
Bond angle deformations	5.9	8.8	24.3
Torsional strain	22.2	17.6	10.0
Non-bonded interactions	23.0	14.6	14.2
Total conformational energy	53.6	46.4	51.4
Energy differences	7.2	0.0	5.0

When trien forms an octahedral complex with unsymmetric bidentates such as amino acid in cis-β co-ordination, two geometric isomers are possible: cis-β_1 with the amino group in a trans-position to a terminal NH_2 group of a trien ligand and cis-β_2 with the NH_2 group in a trans-position to the secondary N atom of the quadridentate. The structures of Δ-(−)$_{589}$-cis-β_1-(R,R) and Δ-cis-β_1-(R,S)-[Co(gly)(trien)]$^{2+}$ have been determined and the observed geometries adequately reproduced by strain energy minimization (Buckingham, Cresswell, Dellaca, Dwyer, Gainsford, Marzilli, Maxwell, Robinson, Sargeson and Turnbull, 1972). The total strain energy difference between Δ-cis-β_1-(R,R) and Δ-cis-β_1-(R,S) isomers is calculated as 3.4 kJ mol^{-1} in favour of the R,R-isomer. The measured difference in ΔH, obtained from the temperature dependence of the equilibrium constant is less than 1.3 kJ mol^{-1} which is in reasonable agreement.

The structures of Δ-cis-β_2-(R,R,S)-[Co(S-pro)(trien)]$^{2+}$ and Λ-cis-β_2-(S,S,S)-[Co(S-pro)(trien)]$^{2+}$ are also known (Freeman and Maxwell, 1970; Freeman, Marzilli and Maxwell, 1970). The observed cis-β_2 co-ordination of S-proline agrees with the prediction that large non-bonded interactions would occur in the alternative cis-β_1 configuration. The major geometrical difference between the Δ-cis-β_2(R,R,S) and Λ-cis-β_2-(S,S,S) isomers consists of the relative orientations of the proline moieties. In the Λ-cis-β_2-(S,S,S) form the pyrrolidine ring is oriented toward the apical trien rings, whereas in the Δ-cis-β_2-(R,R,S) form it is remote from the apical chelate ring. X-ray studies showed that the formation of the Λ-cis-β_2-(S,S,S) isomer is much more reasonable than originally expected from molecular models. The expected severe non-bonded interaction between the bulky pyrrolidine ring and the apical trien ring is alleviated largely by expansion of the bond angles. In fact, the measured free energy difference between these isomers by equlibration on activated charcoal was only 5.4 kJ mol^{-1} in favour of the Δ-cis-β_2-(R,R,S) isomer.

Ethylenediamine-N,N′-diacetic acid $HO_2CCH_2HN-CH_2-CH_2-NHCH_2CO_2H$

In octahedral complexes of the type [Co(edda)(L)], where L represents a bidentate ligand, the quadritentate ligand can co-ordinate in two ways: cis-α and cis-β. When the ligand is unsymmetric R-pn, four isomers are possible for the cis-β form. The structure of one of the four isomers, $\Delta\Delta\Delta\Lambda$-cis,trans-(N−O)-cis-$\beta$-[Co(edda)(R-pn)]$^+$ has been established by X-ray method (Halloran, Caputo, Willet and Legg, 1975).

The ligand edda is a fragment of the larger sexidentate edta. The absolute configuration of the two secondary N atoms is R and S. The glycinato-Co ring is not so strained as in [Co(edta)]$^-$ (see p. 87), since there are only three fused chelate rings in the structure.

4 Quinquedentates

Structures involving only one quinquedentate, 1,4,7,10,13-pentaazatridecane(tetraethylenepentaamine) are known. There are four possible geometric isomers when this ligand forms an octahedral complex (Fig. 4.28).

αα has a mirror plane if the conformations of the chelate rings are ignored and αβ, ββ and β *trans* will have catoptromers. In addition to this, there is the possibility of the existence of diastereoisomeric forms arising from alternative configurations around the secondary N atoms. This leads to two diastereoisomeric forms for the αβ structure and four such forms for the β-*trans* structure (Snow, 1972). Several isomers of [CoCl(tetraen)]$^{2+}$ have been isolated (House and Garner, 1966; Marzilli, 1969). Structures of (+)$_{540}$αβS- and (+)$_{540}$αβR-[CoCl(tetraen)]$^{2+}$ were determined (Snow, 1972), where R and S refer to the configuration about the secondary N atom fusing the chelate rings in the same plane. They are shown in Fig. 4.29. The absolute configurations are both Λ and with reference to the ring numbering of Fig.

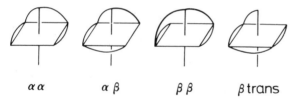

αα   αβ   ββ   β trans

Fig. 4.28 Four possible geometric isomers of [CoX(tetraen)]$^{2+}$

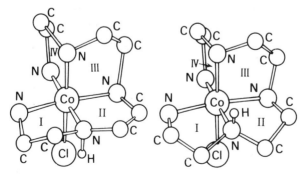

Fig. 4.29. (+)$_{540}$αβR- and (+)$_{540}$αβS-[CoCl(tetraen)]$^{2+}$ (Snow, 1972)

4.29 the rings in sequence I ~ IV have the conformations $\delta\lambda\lambda\delta$ for the $\alpha\beta S$ and $\lambda\delta\lambda\delta$ for the $\alpha\beta R$ isomer. The calculations of the strain energy confirm the experimental result that the $\alpha\beta S$ isomer is more stable by ca. 8.4 kJ mol^{-1}.

5 Sexidentates

Sexidentate complexes of transition metals whose structures are known may be classified into three groups: i) those containing linear ligands, ii) those containing branched ligands and iii) cage compounds.

1,14-Diamino-3,6,9,12-tetraazatetradecane(pentaethylenehexamine)

$$H_2N-CH_2-CH_2-NH-CH_2-CH_2-NH-CH_2-CH_2-NH-CH_2-CH_2-NH-CH_2-CH_2-NH_2$$

This ligand consists of two dien moieties linearly linked by an ethylene group between the primary nitrogen atoms. The complex ion [Co(linpen)]$^{3+}$ can exist in four geometric isomers which are shown in Fig. 4.30. In the isomer A, all the secondary amine groups are arranged in facial positions and can be designated as *ffff*. In B, sets of three N atoms are arranged successively in facial (e, a, b), facial (a, b, c), meridional (b, c, d), and facial (c, d, f) positions.

   Thus starting from one end of the chain, this may be designated *ffmf* (or *fmff*). In the same way C and D can be designated *fmmf* and *mffm* respectively. The above designation, *ffff, ffmf, fmmf* and *mffm* can be written down as acbfed, acbfde, aebcdf and acfbed respectively by using the locant designators. The former notations have an advantage in that they can be used for catoptromers of unknown absolute configuration. When the absolute configurations of the secondary nitrogen atoms are taken into account, the number of co-ordinated isomers increases to eight: I, *ffff-RSSR*, II-1, *ffmf-RSRS;* II-2, *ffmf-RSSS*, III, *fmmf-SRRS;* IV, *fmmf-SSSS;* V, *mffm-RRRR;* VI, *mffm-RRRS;* VII, *mffm-SRRS*, where R and S stand for the ab-

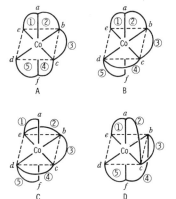

Fig. 4.30. Four possible geometric isomers of [Co(linpen)]$^{3+}$ (Yoshikawa, 1976). Letters a ~ f are the locant designators according to IUPAC nomenclature (IUPAC, 1971)

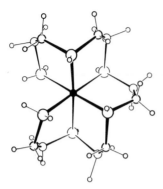

Fig. 4.31. $(+)_{589}[Co(linpen)]^{3+}$ (Sato and Saito, 1975b)

solute configuration of the co-ordinated secondary N atoms and are written in the same order as that of f and m. All these isomers have been separated and resolved into catoptromers (Yoshikawa and Yamasaki, 1973). The characterisation was made by absorption, circular dichroism and pmr spectra. The two conformational isomers II-1 and II-2 can interconvert very rapidly in solution and could not be obtained in pure states. One of the isomers, I, $(+)_{589}[Co(linpen)]^{3+}$ gave crystals suitable for X-ray work as hexacyanocobaltate(III) trihydrate. The structure is shown in Fig. 4.31 (Sato and Saito, 1975b). This is a *ffff* isomer as expected. The absolute configuration is $\Lambda\Lambda\Lambda\Delta$ and the conformations of the chelate rings are $\delta, \lambda, \delta, \lambda$ and δ in turn. The absolute configurations of the secondary N atoms are R, S, S, R. Three of the chelate rings assume unsymmetrical skew conformations and the remaining two are of an eclipsed envelope type. Since all the isomers except I were characterised by physical methods other than X-ray diffraction, strain energy minimization was carried out for all the possible isomers to check the results of characterisation. Table 4.12 lists the final energy terms and formation percentage of the isomers (Yoshikawa, 1976). The isomer, I, contains only the facial arrangement of the co-ordinating N atoms. Thus the bond angle strain is smaller than any others that contain meridional structures. The isomers III to VII, except V, show appreciably large angle bending

Table 4.12. Final energy terms for the isomers of $[Co(linpen)]^{3+}$

Isomer	Bond length deformations	Bond angle deformations	Torsional strain	Non-bonded interactions	Total conformational energy	Formation percentage
I	5.8	15.1	42.9	38.9	102.6 kJ mol^{-1}	9
II-1	5.6	23.3	41.0	35.7	105.6	15
II-2	6.1	22.8	39.7	36.4	105.0	
III	5.6	39.9	34.5	29.9	109.9	1
IV	5.6	34.5	32.7	31.9	104.7	2
V	5.8	22.8	36.1	31.3	95.9	47
VI	6.7	33.1	33.1	32.2	105.0	23
VII	7.4	33.3	33.3	36.1	115.7	3

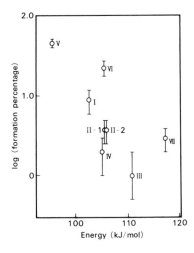

Fig. 4.32. Plot of log (formation percentage) versus minimized energy for the isomers of $[Co(linpen)]^{3+}$

strain, since they contain two mer arrangements of N atoms. The strain in V seems to be alleviated by sacrificing the torsional energy. In Fig. 4.32 logarithms of the observed formation percentage of the isomers are plotted against the minimized total strain energies. As seen from the figure, a roughly linear relationship exists. (The formation percentage of II-1 and II-2 is assumed to be equal and is taken to be one half of the net formation percentage of the two conformers. Furthermore, each formation percentage is divided by 2 for statistical reasons.) As seen from the figure, a roughly linear relationship exists. Such a linear relationship supports the above assignments for the isomers. The relative high abundance of the isomers, VI and VII, when considered with their energies, may suggest a mechanism of formation and/or interconversion in which the very high stability of the isomer V seems to result in their excess formation.

N,N,N',N'-Tetrakis(2'-aminoethyl)-1,2-diaminoethane

The ligand, pentene can function as a sexidentate giving complexes structurally related to those derived from edta. Figure 4.33 shows the complex ion, $(+)_{589}[Co(penten)]^{3+}$ (Muto, Marumo and Saito, 1970). Five five-membered chelate rings are formed in the complex ion. Roughly speaking three chelate rings, A, B and C form a girdle about

$$H_2N-CH_2-CH_2 \diagdown$$
$$\qquad\qquad\qquad N-CH_2-CH_2-N \diagup ^{CH_2-CH_2-NH_2}$$
$$H_2N-CH_2-CH_2 \diagup \qquad\qquad\qquad \diagdown _{CH_2-CH_2-NH_2}$$

the Co atom. Approximately at right angles to the girdle and to one another are the two remaining chelate rings D and E. The co-ordination octahedron is distorted owing to the constraints attending multiple, as well as, fused ring formation. The unique ring A takes unsymmetrical gauche form with δ conformation. Those of the rings B, C and E are λ, whereas that of D is δ. In the rings B and C, the two carbon atoms are

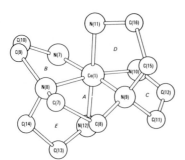

Fig. 4.33. $(+)_{589}[Co(penten)]^{3+}$ (Muto, Marumo and Saito, 1970)

on the same side of the plane formed by the Co and the two N atoms. The ring D takes an envelope conformation. Accordingly, the complex ion, as a whole, does not have a two-fold axis through the midpoint of the C—C bond of the ring A and the Co atom. The absolute configuration can be designated skew chelate pairs $\Lambda\Delta\Lambda$.

(–)N,N,N′,N′-Tetrakis-(2′-aminoethyl)-1,2-diaminopropane

$$H_2N-CH_2-CH_2 \diagdown \qquad\qquad \diagup CH_2-CH_2-NH_2$$
$$N-CH-CH_2-N$$
$$H_2N-CH_2-CH_2 \diagup \quad | \quad \diagdown CH_2-CH_2-NH_2$$
$$CH_3$$

This ligand is a methyl substituted penten and forms an analogous Co(III) complex with $[Co(penten)]^{3+}$. The absolute configuration of $(-)_{589}[Co\{(-)mepenten\}]^{3+}$ has been determined (Kobayashi, Marumo and Saito, 1974). The absolute configuration is $\Delta\Lambda\Delta$ and catoptric to $(+)_{589}[Co(penten)]^{3+}$. The structure of $(-)_{589}[Co\{(-)mepenten\}]^{3+}$ resembles the mirror image of Fig. 4.33 with a substituted methyl group at C(8). However, the conformations of the chelate rings are not exactly the mirror image of $(+)_{589}[Co(penten)]^{3+}$. The unique ring A takes λ conformation with the C—CH$_3$ bond in an equatorial position. The conformations of rings B and C are δ, whereas they are λ in rings D and E. Thus the ring D takes the same conformation as $(+)_{589}[Co(penten)]^{3+}$. This may be due to the packing forces in the crystal lattice. The conformational analysis of these complexes had been carried out before these structures were determined by X-ray diffraction (Gollogly and Hawkins, 1967). The predicted structure can be summarized as follows. i) The methyl group is bonded in equatorial position. ii) Ring A has only one possible conformations(δ for $\Lambda\Delta\Lambda$ absolute configuration and λ for $\Delta\Lambda\Delta$ and rings B and C have a single conformation type which nevertheless permits a small range of conformations with similar bond angles and torsional strains. iii) As to rings D and E, they can exist in either of two distinct conformations which experience similar structural strain. The observed geometries of the two complex ions agree well with these predictions.

Ethylenediaminetetraacetic Acid

$$HO_2C-CH_2 \diagdown N-CH_2-CH_2-N \diagup CH_2-CO_2H$$
$$HO_2C-CH_2 \diagup \qquad \diagdown CH_2-CO_2H$$

The structure of [Co(edta)]⁻ was determined by Wiekliem and Hoard by X-ray diffraction (1959), however, its absolute configuration has only recently been established (Okamoto, Tsukihara, Hidaka and Shimura, 1973). $(+)_{546}$[Co(edta)]⁻ has the absolute configuration ΔΛΔ and the unique five-membered chelate ring takes λ conformation. What makes the difference between this complex and [Co(penten)]³⁺ is the four glycinic rings. The two glycinic rings that form a girdle with the unique ring are strained and take envelope form, while the remaining two out-of-plane chelate rings are less strianed and nearly planar.

Trimethylenediaminetetraacetic Acid

$$HO_2C-CH_2 \diagdown N-CH_2-CH_2-CH_2-N \diagup CH_2-CO_2H$$
$$HO_2C-CH_2 \diagup \qquad \diagdown CH_2-CO_2H$$

The ligand forms an analogous Co(III) complex. The potassium salt, K[Co(trdta)] · 2H₂O resolves spontaneously (Ogino, Takahashi and Tanaka, 1970). $(-)_{546}$[Co(trdta)]⁻ is shown in Fig. 4.34. The complex ion has a rigorous two-fold axis of symmetry through the Co atom and the central C atom of the six-membered chelate ring. The shape of the complex ion is broadly similar to that of [Co(edta)]⁻. The absolute configuration of the entire complex is ΛΔΛ and the six-membered chelate ring assumes a twist-boat form with δ conformation. The two glycinic rings in the plane of the six-membered ring exhibit significant departure from planarity and take envelope form. The other two glycinic rings in the plane nearly perpendicular to the girdle are approximately planar.

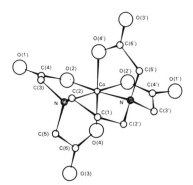

Fig. 4.34. $(-)_{546}$[Co(trdta)]⁻ (Nagao, Marumo and Saito, 1972)

Ethylenediamine-N,N′-diacetic N,N′-di-3-propionic Acid

HOOC–CH₂ 　　　　　　　　　　　 CH₂–CH₂–COOH
　　　　　＼　　　　　　　　　　 ／
　　　　　　N–CH₂–CH₂–N
　　　　　／　　　　　　　　　　 ＼
HOOC–CH₂–CH₂ 　　　　　　　　　 CH₂–COOH

The sexidentate is analogous to edta. The complex ion $(-)_{589}[\text{Cr(eddda)}]^{-}$ takes the absolute configuration $\Lambda\Delta\Lambda$ (Helm, Watson, Radanović and Douglas, 1977). Co-ordination of the Cr(III) ion by eddda creates three five-membered chelate rings and two six-membered chelate rings. The five-membered acetate rings occupy *trans-*axial co-ordination sites, whereas the two-six-membered propionate rings lie in the same plane as the unique five-membered ring (in the girdle plane).

1,3,6,8,10,13,16,19-Octaazabicyclo[6.6.6]eicosane

The octamine ligand acts as a sexidentate and encapsulate metal ions in its cage-shaped skeleton. The $[\text{Co(sep)}]^{3+}$ salt was synthesized by condensation of $[\text{Co(en)}_3]^{3+}$ with formaldehyde and ammonia (Creaser, Harrowfield, Herlt, Sargeson, Springborg, Geue

　　　CH₂–NH–CH₂–CH₂–NH–CH₂
　　／　　　　　　　　　　　　　　＼
N–CH₂–NH–CH₂–CH₂–NH–CH₂–N
　　＼　　　　　　　　　　　　　　／
　　　CH₂–NH–CH₂–CH₂–NH–CH₂

and Snow, 1977). Figure 4.35 presents the $(-)_{589}(S)\text{-}[\text{Co(sep)}]^{3+}$ ion, where S refers to the secondary N atoms. The complex ion has an approximate D_3 symmetry and can be most easily described as a $\Lambda(\delta\delta\delta)\text{-}[\text{Co(en)}_3]^{3+}$ ion with the tris(methylene)ami-no caps added at both ends. The crystal structure analysis revealed that synthesis occurs with retention of the chirality of the $[\text{Co(en)}_3]^{3+}$ ion. Molecular models suggested other possible conformations: a C_3 *lel* and a D_3 conformer has catoptric caps and the D_3 conformers have caps of the same chirality. The strain energy differences of the conformers are 5.9, 5.9 and 0 kJ mol^{-1}. The calculations indicate that the conformer shown in Fig. 4.35 is not the most stable one and it is probably stabilized by hydrogen bonding to lattice Cl^{-} ions. $[\text{Co(sep)}]^{3+}$ can be easily reduced to $[\text{Co(sep)}]^{2+}$ and reoxidizes to $[\text{Co(sep)}]^{3+}$ again with the retention of its chirality. The electron transfer rate is 10^5 fold greater than that for $[\text{Co(en)}_3]^{3+}$. The reason for this difference is not yet well understood.

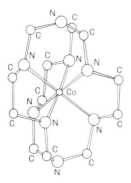

Fig. 4.35. $(-)_{589}(S)\text{-}[\text{Co(sep)}]^{3+}$

Chapter V   Electron-Density Distribution in Transition Metal Complexes

1 Introduction

If we wish to gain insight into the question of what happens when an atom unites with another atom to form a molecule, one of the most direct ways may be to examine the changes which the electronic charge-densities undergo in a process of bond formation. To obtain a measure of this change one can construct a molecule which would result if the atoms making up the molecule were united without perturbing each other, i.e., with spherical charge-density around each nucleus. One can then characterise a chemical bond by the function:

$$\delta\rho(\mathbf{R}) = \rho_M(\mathbf{R}) - \rho_A(\mathbf{R}) \tag{5.1}$$

$\rho_M(\mathbf{R})$ is the electronic charge-density of the molecule M, at some point in space $\mathbf{R}$ and $\rho_A(\mathbf{R})$ is the electronic charge-density at the same point which would occur if the constituent atoms were simply superposed at the molecular equilibrium distance, leaving the molecular geometry unchanged. Thus $\delta\rho(\mathbf{R})$ is positive in regions of the molecule where the charge-density has accumulated and negative where charge-density has moved away. Since net charge is conserved, the integral of $\delta\rho(\mathbf{R})$ over all space is zero. When the molecules assemble together to form crystals, $\rho_M(\mathbf{R})$ is replaced by the electron-density distribution in the unit cell, which is given by a Fourier series, Eq. (2.15) in Chapter II:

$$\rho_M(\mathbf{R}) = \rho_o(xyz) = 1/V \sum |Fo(hkl)| \cos[2\pi(hx + ky + lz) - \alpha(hkl)] \tag{5.2}$$

This can be calculated on the basis of observed structure amplitudes at the final stage of the crystal structure analysis. On the other hand, $\rho_A(\mathbf{R})$ can be given as a Fourier series with calculated structure amplitudes as coefficients:

$$\rho_c(xyz) = 1/V \sum |Fc(hkl)| \cos[2\pi(hx + ky + lz) - \alpha_c(hkl)] \tag{5.3}$$

where

$$F_c(hkl) = \sum_j f_j(hkl) \exp 2\pi i(hx + ky + lz) \tag{5.4}$$

In this equation $f_j(hkl)$ is the usual spherical atomic scattering factor. Thus

$$\rho_A(\mathbf{R}) \equiv \rho_c(xyz)$$

and

$$\delta\rho(\mathbf{R}) \equiv D(xyz) = 1/V \; \Sigma |Fo(hkl) - Fc(hkl)| \; \cos \left[2\pi(hx + ky + lz) - \alpha_c(hkl)\right] \quad (5.5)$$

$D(xyz)$ is called "difference synthesis" which is familiar to X-ray crystallographers. Difference synthesis is used to refine atomic parameters during the process of the structure determination and at the final stage it gives us valuable knowledge of the asphericity of electron density.

The effective charge of an atom in a molecule can be estimated by direct integration of the observed electron-density given by Eq. (5.2) in an appropriate volume, v_j, around each atom:

$$n_j = \int\limits_{v_j} \int \int \rho(xyz) V dx dy dz \qquad (5.6)$$

where

$$\rho(xyz) = 1/V \; \Sigma\Sigma\Sigma |Fo(hkl)| \; \cos \left[2\pi(hx + ky + lz) - \alpha(hkl)\right].$$

For instance, the number of electrons in a sphere of an arbitrary radius R around an atom, $C(R)$, can be easily calculated (Sakurai, 1967). In this way, an effective charge of the central metal atom of a number of complexes has been determined which will be described later. It is difficult to allocate an appropriate volume to each atom, since n_j clearly depends upon the choice of this volume.

Another approach to the problem is the least-squares refinement of the atomic charge-density on the basis of observed structure amplitudes, which is called "electron-population analysis". The atomic scattering factor f_j in Eq. (5.4) is rewritten as a sum of the contribution of core and valence electrons:

$$f_j = f_{j(\text{core})} + p_j f_{j(\text{valence})} \qquad (5.7)$$

where p_j is an adjustable parameter and measures the population of valence electrons. The population of the valence shell can then be determined by least-squares methods after the positional and thermal parameters have been obtained from a conventional least-squares refinement. This was first proposed by Stewart (1970) and called the L-shell projection method. Coppens and his collaborators extended this procedure by refining simultaneously all structural and thermal parameters as well as the populations in the valence shell (1971). In addition the total number of electrons was constrained to be constant in order to keep the crystal neutral during the refinement (the extended L-shell method). One shortcoming of this procedure is that the population parameters depend upon the choice of atomic wave function on the basis of which the core and valence scattering factors are calculated.

The method is further refined and its application will be illustrated later.

Although it has long been recognized that X-ray diffraction determines the electron distribution rather than the atomic positions in a crystal, relatively little attention has been given to a detailed study of charge distribution. This is because most experimental data have not been of sufficient quality for such a study. With the advent of automated four-circle diffractometers and the advance of new data processing techniques both the quantity of data and their quality have improved. Earlier studies were limited to those compounds of lighter elements such as simple organic molecules or metals in the case of heavier elements. Since the 1960's the observation of the effects of bonding electrons has accumulated on organic compounds (Coppens, 1977), however, studies on charge densities in co-ordination compounds have only just been initiated.

As mentioned above the intensity data must be of a good quality. Some attention will now be given to the problems to be faced in obtaining good enough intensity data to warrant charge density analysis. Not all crystals are suitable for an accurate study. The crystal should be stable in air and under X-irradiation. It should give sharp diffraction patterns even at higher Bragg angles. The Fourier series given by Eq. (5.2) is an infinite series and as many intensity data as possible must be collected to minimize termination effects. If the crystal specimen is cooled, the thermal vibration is reduced and the intensity of X-ray reflexion at high Bragg angle is increased which is often too weak to be measured at room temperature.

The intensity of X-ray is measured by scintillation counter. As is well known, when N counts are made in a given time, t, the standard error due to statistical fluctuation is $N^{1/2}$. Thus the diffracted beam intensity (the energy per unit time passing through unit cross section perpendicular to the diffracted beam) is proportional to N/t with a fractional standard error of $N^{-1/2}$. The weak reflexions are measured repeatedly to reduce the standard errors.

The observed structure amplitudes are obtained from a set of measured intensity of reflexions after applying various corrections.

$$E(hkl) = C \, |F(hkl)|^2 \tag{5.8}$$

where $E(hkl)$ is the total energy received by a counter during one Bragg reflexion and C represents various correction factors. In addition to factors concerning the geometry and other conditions of the experiment, the following corrections must be considered:

i) When X-rays pass through a material their intensity is attenuated by absorption. An incident beam entering a crystal as well as the emergent diffracted beam would be attenuated by absorption. This effect can be corrected readily if the shape and size of the crystal specimen and the absorption coefficient are know.

ii) The expression (5.8) will only hold for minute, almost submicroscopic crystals of the order of 10^{-5} cm in diameter. If ordinary macroscopic crystals (usually with a dimension of about 0.2 mm) are perfect throughout their volume, interaction of the incident beam with the diffracted beam in the crystal results in a destructive interference and the total energy of reflexions is less than that indicated by Eq. (5.8) and is proportional to $|F(hkl)|$ (not its square). It is found however that Eq. (5.8) is in fact applicable, to many single crystal specimens providing allowance is made for this effect, called primary extinction. It is believed that the crystal behaves as if it consisted of a number of blocks, each block being a perfect crystal but adjacent blocks not accurately fitting together (mosaic structure). This imperfection is perhaps connected with crystal dislocations accompanying crystal growth. The intensities of the strongest reflexions are reduced sometimes by a "secondary extinction effect". The top layer of a crystal, i.e. the part nearest to the incident beam reflects away an appreciable proportion of the primary beam, thus in effect partially shielding the lower layers of the crystal: the strongest reflexions are experimentally less strong than they should be in comparison with the weaker reflexions. The secondary extinction effect depends upon the distribution of the shape and size of the mosaic blocks and their mode of alignment. The extinction effects are not yet fully understood and the theory is based on a rather artificial mosaic block model. Methods for correcting extinction effects have been proposed (for example, see Coppens and Hamilton, 1970).

iii) Atoms in crystals undergo thermal vibrations with very low frequencies compared to the sort of transmission time of X-rays in crystals. For this reason we may imagine that, at any instant of time, the diffraction pattern produced is that of a "frozen crystal" in which all the atoms are stationary and displaced randomly at some distance from their equilibrium positions. The total intensity of reflexion measured over any long period of time is a time average of patterns which are obtained at successive instants. Thus thermal vibration has the effect of smoothing out the charge-density distribution. An obvious way to reduce this effect is to collect the intensity data at low temperatures, where the thermal vibrations are suppressed. The second way of avoiding this effect is to use crystals in which thermal vibration amplitudes are very small even at room temperatures. Examples of such crystals are double oxides such as spinel and some actual examples will be described later.

iv) The atomic positions are usually obtained by the least-squares fit between the observed and calculated X-ray structure amplitudes. Calculation of the structure amplitudes is based on the spherical atomic scattering factor and the deformation of charge-density from spherical distribution is neglected. Thus the atomic parameters will be biased where atoms are placed in an asymmetric environment, like terminal oxygen atoms, nitrogen and especially hydrogen atoms. The atomic parameters obtained from neutron diffraction data are, however, quite free from the bias mentioned above, since neutron interacts with atomic nuclei rather than with the electron clouds around the nuclei. Accordingly, the atomic parameters based on neutron diffraction data are used for the calculation of ρ_c. Such syntheses are called X—N syntheses and frequently used. It is, however, possible to obtain good approximation to X—N synthesis on the basis of X-ray data, without recourse to expensive neutron diffraction experiments. The theory of X-ray scattering indicates that the intensity of high angle reflexions is almost exclusively determined by the core electrons of the atoms and relatively unaffected by the valence density. Thus the reasonable way of obtaining the deformation of electron-density due to bonding exclusively from X-ray data would be to employ the atomic parameters obtained on the basis of high angle X-ray diffraction data for the calculation of $\rho_c(xyz)$. The results obtained for the X-ray and neutron studies on identical transition-metal complexes indicated that these two difference densities resemble each other quite closely. The same consideration is applicable to electron population analysis.

2 Earlier Work

Prior to discussion of the recent advances gained with regard to transition-metal complexes, some important results obtained by earlier workers will be described in order to show how the electron-density distribution changes according to the nature of the bonds in crystals. Figure 5.1 is the electron-density distribution of sodium chloride (Witte and Wölfel, 1955). The contour lines are circular, indicating that the electron clouds around the atomic nuclei are spherical. Between the region of Cl and Na, the electron-density is zero. Electron clouds are not continuous. The number of electrons was estimated by integration of $\rho(xyz)$ over a reasonable volume. It was shown that the number of electrons around Na is 10.1 and that around Cl is 17.8. The atomic numbers of Na and Cl are 11 and 17 respectively. The result indicates that the sodium atom is in the state Na^+ and the chlorine atom is in the state Cl^-. Thus sodium chloride is composed of Na^+ and Cl^- ions, which may be condisered as incompressible spheres and are usually only slightly polarised by the ions of op-posite charge. This is a typical ionic crystal. The crystals of NaCl are hard and have a high melting point, because the ions are held together by electrostatic forces. The ions cannot move in the crystal lattice, thus the crystal is an insulator. When NaCl is dissolved in water or molten, the ions can move and the aqueous solution or the melt is conductive.

   With regard to results obtained for metals, aluminium for instance, crystallizes in a cubic face-centred lattice, i.e. cubic closest packing of spheres. The results of ac-curate determinations of the electron-densities can be summarised as follows

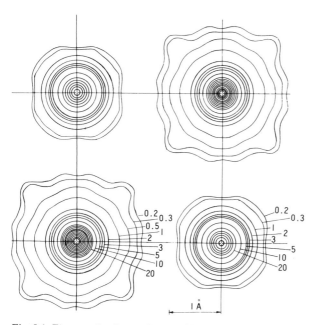

Fig. 5.1. Electron-density section in a NaCl crystal through the plane xyO (Witte and Wölfel, 1955). Numerals indicate e Å^{-3}

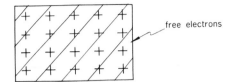

Fig. 5.2. Free electron model of a metal

(Bensch, Witte and Wölfel, 1955; Brill, Hermann and Peters, 1944; Bensch, Witte and Wölfel, 1954): Unlike NaCl, the electron-density does not reach the value zero anywhere in the lattice. The background density is $0.18e$ Å^{-3}. This density corresponds to about 3 electrons per aluminium atom uniformly distributed throughout the lattice. By integration over the spherical atom a total of 10.2 electrons per aluminium atom was obtained. This result is close to the 10 electrons for the Al^{3+} ion. Such electron-distribution in a metal agrees well with the suggestion made by Lorentz that a metal consists of an array of cations in a sea of free electrons (Fig. 5.2). "Free electrons" can move about from atom to atom, thus accounting for electrical conductivity, and can respond to electromagnetic waves, thus accounting for the reflection of light.

Figure 5.3 illustrates a diamond structure, in which a carbon atom is tetrahedrally bonded to four others. They are bonded covalently by electron pairs which occupy localised molecular orbitals formed by overlapping sp^3 hybrids. The four carbon atoms A, B, C and D are on one plane. Figure 5.4 shows the section of the difference synthesis through this plane. There is a positive peak with height 0.51 e Å^{-3} between the carbon atoms, indicating that the electron density has moved to the centre of the bond. This peak is due to σ-bonding electrons (Göttlicher and Wölfel, 1959).

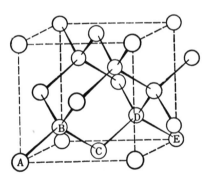

Fig. 5.3. Diamond structure

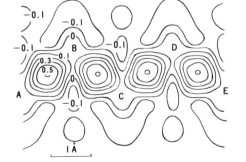

Fig. 5.4. The section of the difference synthesis of a diamond crystal through the plane of carbon atoms formed by A, B, C and D shown in Fig. 5.3 (Göttlicher and Wölfel, 1959)

3 Charge-Density Distribution in Transition-Metal Complexes: Preamble

Two important features are revealed by an accurate study of charge-density distribution in transition-metal complexes: the nature of the metal-ligand bond and the behavior of d-electrons in the ligand field.

i) Metal-Ligand Bonds

A co-ordination bond is formed by sharing an electron-pair between a metal atom and a ligand. The electron pair is donated by the ligating atoms. One important factor that affects the stability of the bond is the partial ionic character. Another factor may be the multiple bond character of the metal ligand bonds. Let us consider hexaamminecobalt(III) ion, $[Co(NH_3)_6]^{3+}$. If the Co-N bond were ionic, the elec-

$$
\begin{array}{cc}
\begin{array}{c}
H_3 \\
N \\
H_3N: \quad \quad :NH_3 \\
\quad Co^{3+} \\
H_3N: \quad \quad :NH_3 \\
\ddot{N} \\
H_3
\end{array}
&
\begin{array}{c}
H_3 \\
\overset{+}{N} \\
H_3\overset{+}{N}: \quad \quad : \overset{+}{N}H_3 \\
\quad \overset{3-}{Co} \\
H_3\overset{+}{N}: \quad \quad : \overset{+}{N}H_3 \\
\overset{+}{N} \\
H_3
\end{array}
\end{array}
$$

tric charge 3+ would be located on the cobalt atom, and if they are extremely covalent in nature, the cobalt atom would have the charge 3− and each nitrogen atom the charge 1+. Such charge distribution may be unstable and the bond may in fact, have partial ionic character.

ii) d-Electron Distribution in a Ligand Field

When a transition metal ion is surrounded octahedrally by six ligating atoms, the five-fold degenerate 3d levels have no more equal energy but split into an upper doublet e_g and a lower triplet t_{2g}. Figure 5.5 compares the environments of $d_{x^2-y^2}$ and d_{xy} orbitals placed in an octahedral arrangement of six ligating atoms. By symmetry, the ligating atom along the z axis influences the $d_{x^2-y^2}$ and d_{xy} orbitals to the same extent, but the situation for those on the x and y axis is clearly different. The electrons in the $d_{x^2-y^2}$ orbital are repelled more strongly by the ligating atoms than those in the d_{xy} orbital. Similar consideration for other orbitals leads to the conclusion mentioned above and the resulting energy level schemes are shown in Fig. 5.6 (a). Next consider the case of a transition-metal ion placed in a tetrahedral environment (Fig. 5.6 (b)). Comparison of Figs. 5.5 and 5.7 at once reveals that the d_{xy}, d_{yz} and d_{zx} orbitals are destabilised to a greater extent than the d_{z^2} and $d_{x^2-y^2}$ orbitals, leading to the energy level schemes shown in Fig. 5.6 (b).

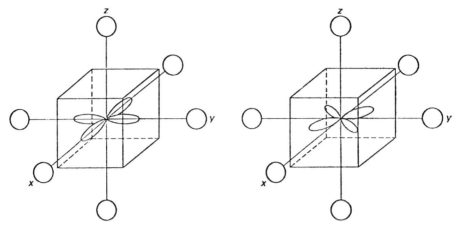

Fig. 5.5. The $d_{x^2-y^2}$ and d_{xy} orbitals in an octahedral environment

$$\underline{\underline{d_{x^2-y^2},\ d_{z^2}}}\quad e_g$$

$$\underline{\underline{d_{xy},d_{yz},d_{zx}}}\quad t_{2g}$$

$$\underline{\underline{d_{xy},d_{yz},\ d_{zx}}}\quad t_2$$

$$\underline{\underline{d_{x^2-y^2},\ d_{z^2}}}\quad e$$

(a)   (b)

Fig. 5.6. Energy level schemes for **(a)** octahedral and **(b)** tetrahedral co-ordination

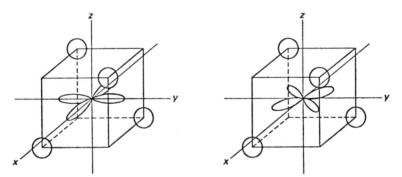

Fig. 5.7. The $d_{x^2-y^2}$ and d_{xy} orbitals in a tetrahedral environment

In the language of molecular orbital theory, a more realistic physical picture of the interaction between ligand and metal is based on the recognition that the atomic orbitals necessarily become mixed to form molecular orbitals when the nuclei approach to within a few Ångstroms of one another. It is immediately clear that there can be no net overlap between a t_{2g} [13] orbital on the metal and a σ orbital of a ligand lying on one of the co-ordinate axes (Fig. 5.8).

13 Lower case letters e and t are used when referring to energy states or orbitals whereas upper case E and T are reserved for symmetry species.

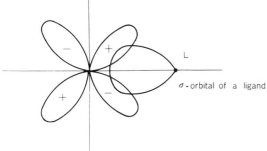

σ-orbital of a ligand

a metal atom

Fig. 5.8. Zero overlap of a t_{2g} orbital with a σ orbital on one of the co-ordinate axes

Now we consider the charge-density distribution of t_{2g} and e_g orbitals. Neglecting the radial part, the angular part of the five d orbitals are:

$d_{xy} = (15/16\pi)^{1/2} \sin^2 \theta \sin 2\phi$
$d_{yz} = (15/4\pi)^{1/2} \sin \theta \cos \theta \sin \phi$
$d_{zx} = (15/4\pi)^{1/2} \sin \theta \cos \theta \cos \phi$
$d_{z^2} = (5/16\pi)^{1/2} (3 \cos^2 \theta - 1)$
$d_{x^2-y^2} = (15/16\pi)^{1/2} \sin^2 \theta \cos 2\phi$

By squaring and adding, the total charge density of t_{2g} and e_g orbitals are obtained as follows:

t_{2g} orbital charge density $= d_{xy}^2 + d_{yz}^2 + d_{zx}^2$

$$= (15/16\pi \sin^2 \theta (4 \cos^2 \theta + \sin^2 \theta \sin^2 2\varphi) \qquad (5.9)$$

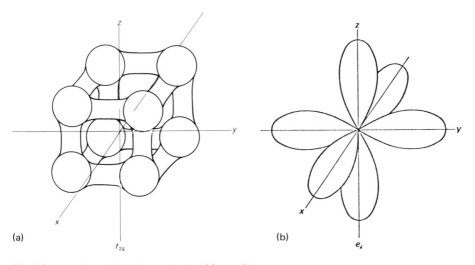

(a) t_{2g}

(b) e_g

Fig. 5.9. t_{2g} and e_g orbital charge density (a) t_{2g}, (b) e_g

e_g = orbital charge density = $d_{z^2} + d_{x^2-y^2}$

$$= (5/16\pi)(3\sin^4\theta\cos^2 2\varphi + 9\cos^4\theta - 6\cos^2\theta + 1) \qquad (5.10)$$

By differentiation with respect to θ and φ, one can easily verify that t_{2g} orbital charge density has eight lobes on the diagonals of a cube and that e_g has six lobes pointing towards the faces of a cube as shown in Fig. 5.9.

4 Charge-Density Distribution in Transition-Metal Complexes

A. [Co(NH₃)₆][Co(CN)₆]

The crystal contains two types of the most familiar and fundamental complex ions of cobalt(III). Figure 5.10 shows the packing diagram of the complex salt. The crystal is ionic and the complex cations and anions are arranged like the ions in caesium chloride. The ions are held together by hydrogen bonds between N—H and the nitrogen atom of a cyanide group. They are shown by dotted lines in Fig. 5.10. The number of electrons in a sphere of radius R around the cobalt atom was computed by direct integration of the observed electron-density. The standard deviation of the observed electron-density at the general position is $0.06\ e\ \text{Å}^{-3}$ and that at the metal site is $0.2\ e\ \text{Å}^{-3}$. The result is shown in Fig. 5.11. The observed curve for [Co-(NH₃)₆]³⁺ agrees well with the calculated values obtained by means of the Thomas-Fermi method (Kamimura, Koide, Sugano and Tanabe, 1958). The original result is modified for the effect of thermal vibration, the root-mean squared amplitude of vibration being $0.12\ \text{Å}$. The number of electrons around the cobalt atom in [Co-(NH₃)₆]³⁺ in a sphere of radius $1.22\ \text{Å}$ (covalent radius of cobalt) is 26.3 ± 0.3. In Table 5.1 the number of electrons determined from $C(R)$ for the two complex ions and from the least-squares fitting of the scattering factor curves is compared with the calculated values. The number of electrons around the cobalt atom in the complex

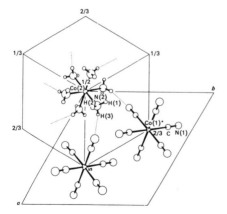

Fig. 5.10. A packing diagram of [Co(NH₃)₆][Co(CN)₆] (Iwata and Saito, 1973)

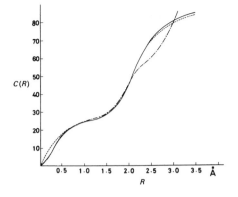

Fig. 5.11. Number of electrons in a sphere of radius R
———— observed value for $[Co(NH_3)_6]^{3+}$
– – – calculated value for $[Co(NH_3)_6]^{3+}$
—·—·— observed value for $[Co(CN)_6]^{3-}$
(Iwata and Saito, 1973)

cation, $[Co(NH_3)_6]^{3+}$, is in good agreement with the calculated values. The nature of the electron cloud about a nucleus, however, makes it difficult to define the size of an atom. Thus the covalent radius of an atom has no precise physical significance. As will be described below, there is a peak due to bonding electrons between the metal and the ligating atom. Accordingly, if the covalent radius is replaced by the distance from the nucleus to the bonding electron-density peak, the concept of "charge density of the central metal atom" will be much clearer. Table 5.2 lists the effective charge of the central metal atoms in $[Co(NH_3)_6]^{3+}$ and $[Co(CN)_6]^{3-}$ defined in this way, together with those obtained for other complexes. The results listed in Tables 5.1 and 5.2 indicate that the central metal atom is largely neutralised by electron donation from the ligating atoms. The effective charge on the metal atom is about +0.5 e in $[Co(NH_3)_6]^{3+}$. More detailed electron population analysis revealed that the effective charge on NH_3 is about +0.5 e (Iwata, 1977). This means that the charge 3+ of the complex ion, $[Co(NH_3)_6]^{3+}$ is distributed on six ammonia groups (mainly on eighteen hydrogen atoms). Such charge distribution is consistent with that on a macroscopic body: when one gives electric charge to a body, the charge distributes on the surface. These results indicate that Pauling's electroneutrality rule holds for transition metal complexes (1960). The ionicity of the metal ligand bond is thus ca. 50%.

Table 5.1. Number of electrons around the cobalt atom

	Direct integration of electron density	Least-squares fitting of scattering factor curves	Kalman and Richardson (1971)	Kamimura et al. (1958)
$[Co(NH_3)_6]^{3+}$	26.3 ± 0.3[a]	26.2 ± 0.1	26.4	25.9
$[Co(CN)_6]^{3+}$	26.8 ± 0.3	26.6 ± 0.1	—	—

a   Within a sphere of radius 1.22 Å

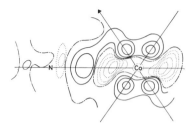

Fig. 5.12. Section of final difference synthesis through a Co−N bond and a three-fold axis of $[Co(NH_3)_6]^{3+}$ in crystals of $[Co(NH_3)_6][Co(CN)_6]$ (Iwata and Saito, 1973). An arrow indicates the three-fold axis. The solid contours are at intervals of 0.1 e A^{-3}. Negative contours are dotted, zero being chain dotted

Figure 5.12 shows a section of the final difference synthesis containing a three-fold axis of the complex ion and a Co−N bond. A maximum appears between cobalt and nitrogen atoms while there are minima on either side of the bond. The height of the positive region between the cobalt and nitrogen atoms is approximately 0.30 e A^{-3} and the minima range from −0.20 to −0.40 e A^{-3}. Such an arrangement of maxima and minima suggests that electron density has moved to the centre of the bond due to donation of the lone pair electrons from an ammonia molecule. The four maxima around the cobalt atom are due to the asphericity of 3d electron distribution in the octahedral ligand field, which will be discussed below. The corresponding section of the difference synthesis of $[Co(CN)_6]^{3-}$ is presented in Fig. 5.13. The arrangement of the maxima and minima in the section is much the same as that observed for $[Co(NH_3)_6]^{3+}$. However, the residual electron densities between Co and C and between C and N are much more elongated perpendicularly to the bonds than is observed for the Co−N bond. The difference Fourier sections bisecting the bonds are presented in Fig. 5.14. The elongations may be ascribed to $d_\pi - p_\pi$ and $p_\pi - p_\pi$ overlap respectively. The section of the C−N bond shows a distribution like a torus around the bond, the maximum density of the torus being about 0.10 e A^{-3} with a radius of about 0.7 Å. It can be seen that the π-electron density calculation on the basis of a wave function of the form $(2p_x) = Nx \exp(-cr/2)$ shows a maximum of

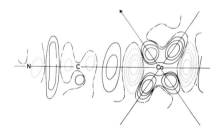

Fig. 5.13. Section of the final difference Fourier synthesis through a Co−C−N bond and a three-fold axis of $[Co(CN)_6]^{3-}$. An arrow indicates the three-fold axis. The scale is the same as in Fig. 5.12 (Iwata and Saito, 1973)

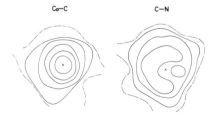

Fig. 5.14. Difference Fourier sections bisecting the bond. The contours are at intervals of 0.05 e A^{-3}

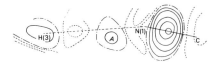

Fig. 5.15. Section of difference Fourier synthesis through a C−N bond and a hydrogen atom hydrogen bonded to the nitrogen atom. The contours are at intervals of 0.05 e Å^{-3}. Negative contours are broken, zero being chain-dotted (Iwata and Saito, 1973)

0.12 e Å^{-3} at 0.70 Å from the midpoint of the carbon and nitrogen atoms (Coulson, 1961). Similar features have been observed in the residual electron-density maps of other organic compounds (for example, Mathiews and Stucky, 1971; Tsuchiya, Marumo and Saito, 1972; Tsuchiya, Marumo and Saito, 1973). Another interesting feature is observed in the section through a cyano group and a hydrogen atom bonded to the nitrogen atom of the cyano group, which is shown in Fig. 5.15. A small peak indicated as A lies at 0.75 Å from N(1) and slightly off the line through the C−N(1) bond. Though not significant, this small peak appears to be due to the localized nitrogen lone pair directed towards H(3).

B. $[Co(NH_3)_6][Cr(CN)_6]$

Figure 5.16 shows sections of the difference synthesis of $[Co(NH_3)_6][Cr(CN)_6]$ at 80 K. (a) and (b) correspond to Figs. 5.12 and Fig. 5.13 respectively. At 80 K, amplitudes of thermal vibrations of atoms were reduced to about two-thirds of those at room temperature. The bonding electrons are clearly seen in the sections through the Co−NH$_3$, Cr−C and C−N bonds. In particular the section of the C−N triple bond shows two circular sections of a torus shaped charge distribution of the $p_\pi - p_\pi$ bond.

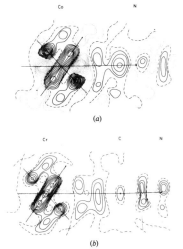

Fig. 5.16. Sections of the difference map of [Co (NH$_3$)$_6$][Cr(CN)$_6$] at 80 K. (a) a section through a Co−N bond and the three-fold axis (▲); (b) a section through a Cr−CN bond and the three-fold axis. Contours are at 0.1 e Å^{-3} intervals; negative contours are drawn by dotted lines, zero being chained. The crosses indicate atom positions (Iwata, 1977)

C. Asphericity of d-Electron Distribution in [Co(NH₃)₆][Co(CN)₆] and [Co(NH₃)₆][Cr(CN)₆]

The aspherical charge distribution of transition metal atoms in complexes was first detected by Iwata and Saito for $[Co(NH_3)_6][Co(CN)_6]$. The residual electron density map shown in Fig. 5.12 corresponds to a section through a face diagonal of the cube formed by eight residual peaks due to six 3d electrons in the non-bonding t_{2g} orbital of $[Co(NH_3)_6]^{3+}$ (See Fig. 5.9). There are four peaks at 0.45 Å from the cobalt nucleus with a peak height of 0.3 e Å^{-3}. In fact, eight peaks are arranged approximately at the corners of a cube around the cobalt nucleus as predicted theoretically. Similar features are observed for $[Co(CN)_6]^{3-}$ (Fig. 5.13). The positions of the maximum density of t_{2g} orbitals for a free cobalt atom lie at 0.35 Å from the nucleus (Clementi, 1965). This distance generally increases when the atom is placed in a ligand field. For instance, in NiO this distance increases to 0.6 Å (Watson and Freeman, 1960). Shull and Yamada showed by neutron diffraction that the maxima due to 3d electrons are located at 0.5 Å from the nucleus in Fe crystals (1962).

The asphericity of d electron distribution observed for $[Co(NH_3)_6]^{3+}$ and $[Cr(CN)_6]^{3-}$ at 80 K [Fig. 5.16 (a) and (b)] differs considerably from the asphericity at room temperature (Figs. 5.12 and 5.13). The reason is that the symmetry of the complex ions (as well as the site symmetry) is not octahedral O_h but, in fact, trigonal C_{3i}, although the deviation from regular octahedral symmetry is very small: NCoN' $= 89.50 \pm 0.03°$, CCrC' $= 90.16 \pm 0.04°$, where the primed atoms are related to those without primes by a three-fold axis. Under such trigonal field the t_{2g} orbital further splits into a lower singlet a_g and a doublet e_g. This situation is more clearly reflected in the difference synthesis at 80 K than at room temperature, since the asphericity is blurred owing to the thermal vibration of atoms. In Fig. 5.16 (a) and (b) the peaks on the three-fold axis are higher and the maxima are closer to the metal atoms. On closer examination of Figs. 5.12 and 5.16 one can see that even at room temperature the distribution of the eight peaks is not regular octahedral but distorted trigonally to a small extent: the peaks on the three-fold axis differ slightly from other peaks. Further detailed analysis of the charge distribution has been carried out (Iwata, 1977).

D. K₂Na[Co(NO₂)₆]

The compound is non-stoichiometric. The ideal empirical formula is $K_2Na[Co(NO_2)_6]$. The crystal structure is cubic and cryolite in type. Na$^+$ and $[Co(NO_2)_6]^{3-}$ ions are arranged like the ions in sodium chloride and the potassium ions are located at the centre of a cube formed by four Na$^+$ and four $[Co(NO_2)_6]^{3-}$ alternately occupying eight corners of the cube. The crystal specimen used for this work has the composition: $K_{1.64}Na_{1.36}[Co(NO_2)_6]$ and the excess sodium ions were found to occupy the K-site randomly (Ohba, Toriumi, Sato and Saito, 1978). Six nitro groups co-ordinate to Co(III) octahedrally with a Co—N distance of 1.952 Å. Fig. 5.17 (a) shows a section of the final difference synthesis through the cobalt atom and the two three-fold axes of rotation, which corresponds to Fig. 5.12 and Fig. 5.17 (b) is a section

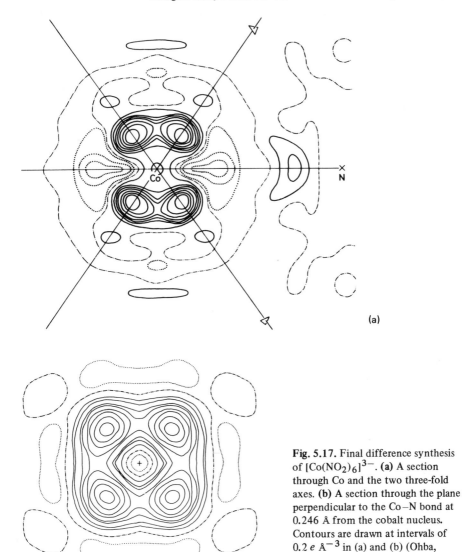

Fig. 5.17. Final difference synthesis of $[Co(NO_2)_6]^{3-}$. **(a)** A section through Co and the two three-fold axes. **(b)** A section through the plane perpendicular to the Co—N bond at 0.246 Å from the cobalt nucleus. Contours are drawn at intervals of 0.2 e Å^{-3} in (a) and (b) (Ohba, Toriumi, Sato and Saito, 1978)

through the plane perpendicular to the Co—N bond and at 0.246 Å from the cobalt nucleus. This is a section through the cube face formed by the residuel electron-density peaks. Eight peaks with heights of 1.7 (0.1)[14] e Å^{-3} are arranged at the eight corners of a cube at 0.43 Å from the cobalt nucleus. This feature is exactly the same as that predicted by ligand field theory for 3d electrons in non-bonding orbitals. A peak due to bonding electrons is located on the Co—N bond at 1.43 Å from the cobalt atom. A simple molecular orbital model for the NO_2^- group can be constructed

---

14 Number in parentheses stands for the estimated standard deviation.

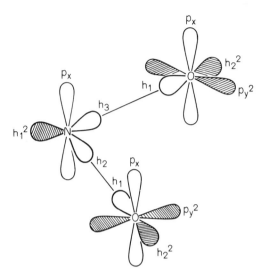

Fig. 5.18. A simple molecular orbital model of NO_2^-

as follows: The NO_2^- group is planar with the nitrogen atom at the centre and has a ONO angle of about $120°$. The valence state of N must be described in terms of three similar hybrid orbitals pointing towards the corners of a regular triangle. Such orbitals of the N atom can be formed by mixing the 2s atomic orbital and two 2p atomic orbitals, say $2p_y$ and $2p_z$; the hybrid orbitals lie in the yz plane of the latter and are precisely equivalent (Fig. 5.18). If these hybrids are denoted by h_1, h_2 and h_3, the appropriate nitrogen valence state must be $N(1s^2, h_1^2, h_2^1, h_3^1, 2p_x^1)$. $N(h_1^2)$ forms a lone pair and its lobe is oriented toward the cobalt atom to form a co-ordination bond. The relevant orbital on the O atom is the digonal hybrid formed by 2s and $2p_z$ (z referring to the N–O axis) and the valence state is $O(1s^2, h_1^1, h_2^2, 2p_x^1, 2p_y^2)$, where h_1 and h_2 denote the two lobes of the digonal hybrids. $O(h_2^2)$ forms a lone pair located at the rear of the O atom with respect to the N atom. $O(2p_y^2)$ is also a lone pair and the lobes are perpendicular to the N–O bond and in the plane of the nitro group. The $N(h_2^1)$ orbital overlaps with $O(h_1^1)$ to form a σ-bond and $N(2p_x^1)$ and $O(2p_x^1)$ overlap to form a π-bond. An excess electron is delocalised on both $O(2p_x)$ orbitals:

$$:N\diagdown\begin{matrix}=O\\ \\O^-\end{matrix} \quad \rightleftarrows \quad :N\diagdown\begin{matrix}O^-\\ \\ =O\end{matrix}$$

Figure 5.19 shows a section of the difference synthesis through the plane of a NO_2 group. The distribution of the residual electron density agrees well with the simple picture outlined above. A bonding peak with height 0.32 (0.1) e $Å^{-3}$ is observed between N and O atoms and two peaks are located on a line through the O atom and perpendicular to the N–O bond. They may be ascribed to $2p_y^2$ lone pair electrons; however, the peak heights are unequal; a peak in the ONO angle of 0.4 (0.1) e $Å^{-3}$ is much higher than the other one outside it 0.2 (0.1) e $Å^{-3}$. This is

Fig. 5.19. A section of the final difference synthesis through a nitro group. Contours are drawn at intervals of $0.1\ e\ \text{Å}^{-3}$ (Ohba, Toriumi, Sato and Saito, 1978)

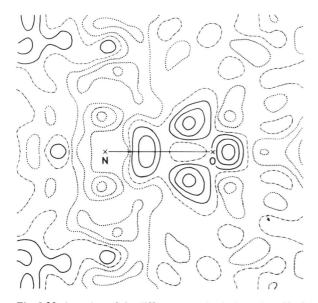

Fig. 5.20. A section of the difference synthesis through an N–O bond and perpendicular to the NO_2 group. Contours are drawn at intervals of $0.1\ e\ \text{Å}^{-3}$ (Ohba, Toriumi, Sato and Saito, 1978)

because a sodium ion lies on the Co–N axis at 5.12 Å from the Co atom and the electrostatic interaction between Na^+ and the lone pair electrons stabilises the lobe in the ONO angle.

A peak near the O atom on N–O axis resembles lone-pair electrons in one of the two digonal hybrids, however, its significance is low, since such a peak near the O atom is seriously affected by the errors of the positional parameters of the O atom. Fig. 5.20 shows a section of an N–O bond perpendicular to the plane of the NO_2 group. The bonding electron in the N–O bond is largely extended perpendicularly to the plane of the NO_2 group, giving two peaks of 0.3 (0.1) e Å^{-3} at 0.31 Å above and below the plane of the group. This feature may indicate that the N–O bond is π-bond in character. Charge density distribution around the cobalt atom is very similar to that of $[Co(NH_3)_6]^{3+}$. The number of electrons within a sphere of radius 1.22 Å is 26.3 (0.1) e. Effective charge on each atom was estimated as follows: N, $-0.07\ e$; O, -0.22 (0.02) e.

E. Charge Distribution in Spinels

The asphericity of d electron distribution in a ligand field can be more clearly observed in spinel crystals in which thermal vibration amplitudes of the constituent atoms is small (r.m.s. amplitude is 0.07 Å on the average, while it is about 0.1 Å in transition metal complexes). The spinels are minerals with the empirical formula AB_2O_4. There are 2:3 spinels containing A^{2+} and B^{3+} ions, and 4:2 spinels containing A^{4+} and B^{2+} ions. In a normal 2:3 spinel structure, the array of oxide ions forms a cubic close packed structure and each B^{3+} ion is surrounded octahedrally by six oxide ions and each A^{2+} ion is tetrahedrally surrounded by four oxide ions. The asphericity of d electron distribution was observed first in a 4:2 spinel, γ-

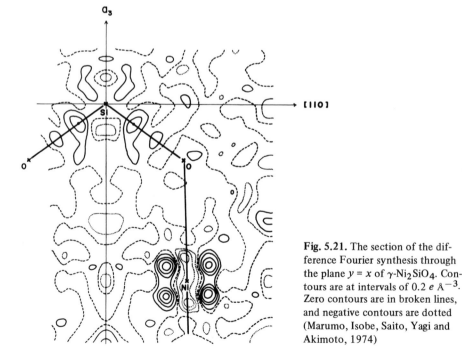

Fig. 5.21. The section of the difference Fourier synthesis through the plane $y = x$ of γ-Ni_2SiO_4. Contours are at intervals of 0.2 e Å^{-3}. Zero contours are in broken lines, and negative contours are dotted (Marumo, Isobe, Saito, Yagi and Akimoto, 1974)

Ni_2SiO_4 (Marumo, Isobe, Saito, Yagi and Akimoto, 1974). The compound has a strictly normal spinel structure, in which the Ni^{2+} ion is surrounded octahedrally by six oxygen atoms and Si atoms have tetrahedral sites. A section of the difference Fourier map through the plane $y = x$ is shown in Fig. 5.21. All the crystallographically independent atoms and chemical bonds in this structure appear in this section. One Ni-O bond is shown in the figure. A remarkable feature of the map is the existence of four salient peaks 0.46 Å from the nickel atom. In the three-dimensional maps, there are eight such peaks around the nickel atom, of which two are crystallographically independent. The heights of the peaks on the three-fold rotation axis of the crystal are $0.9 e$ Å^{-3} and the other $1.0 e$ Å^{-3}. These eight peaks are disposed at the eight corners of a cube surrounding the nickel atom. Six negative peaks of height $-0.6 e$ Å^{-3} are located at about 0.4 Å on each Ni-O bond from the nickel atom. The negative peaks are at the apices of an octahedron. Since the Ni^{2+} ion is placed in a nearly octahedral field, six out of the eight 3d electrons of the ion are in the t_{2g} orbitals and the other two are in the e_g orbitals in the ground state. Then excess electron density is expected in t_{2g} orbitals and deficiency for e_g orbitals, on the difference map which shows the deviation of the electron density from spherical distribution. The observed features in the difference map reflect such aspherical distribution of d electrons. A nickel atom in an octahedral site might vibrate with a larger amplitude along the body-diagonals perpendicularly to the octahedral face than along the Ni-O bonds. Such thermal vibration would also give rise to the feature mentioned above. This effect might not be serious in the case of ionic crystals like spinels with rather small thermal vibrations (r.m.s. amplitude of the Ni atom being 0.066 Å). Because of such small thermal vibrations, the asphericity of charge

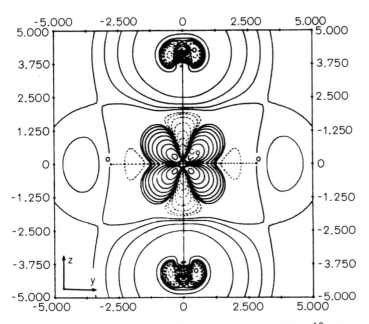

Fig. 5.22. A section of calculated deformation density for $[CoO_6]^{10-}$. First contours are respectively $+0,0025 e$ (a.u.)$^{-3}$. Neighbouring contours differ by a factor of 2 (Johansen, 1976)

distribution can be observed more markedly than in the case of Werner complexes. The same feature is observed in the difference Fourier synthesis of isostructural γ-Co_2SiO_4 (Co^{2+} : $3d^7$) (Marumo, Isobe and Akimoto, 1977).

On the other hand, theoretical calculation was made for the asphericity of d-electron charge density by using ab initio Hartree-Fock and configuration interaction methods (Johansen, 1976). Fig. 5.22 shows the deformation density for $[CoO_6]^{10-}$ in the $^4T_{1g}$ ground state, the calculation having included configuration interaction. The result is quite similar to those shown in Figs. 5.12, 5.13 and 5.21. The calculated map shown in Fig. 5.22 is drawn quite differently from that obtained experimentally. Firstly the density at each contour is double that of its neighbour so that the contour intervals are not equal. Secondly the density contours are drawn in units of electrons per cubic atomic unit, where $1\ e\ (a.u.)^{-3} = 6.74876\ e\ Å^{-3}$. Therefore care must be taken when making comparisons. The pronounced peaks close to the nucleus are at distances of 0.25 Å, with heights of 4 e $Å^{-3}$. The peaks are located closer to the nucleus and higher than those detected experimentally. The main reason for this discrepancy may be the inadequate description of the thermal motion in the theoretical treatment. Fig. 5.23 shows a section of the difference map of γ-Fe_2SiO_4 through the plane corresponding to that of γ-Ni_2SiO_4 (Marumo, Isobe and Akimoto, 1977). In γ-Fe_2SiO_4 the Fe^{2+} ion is in the high-spin state with six 3d electrons. The cation is octahedrally surrounded by six oxide ions; the co-ordination octahedron is not regular but flattened along its three-fold axis. Because of this trigonal deformation the triply degenerate t_{2g} level is split into one singlet a_g and one two-fold degenerate level e_g. From the flattened shape of the co-ordination octahedron, the singlet a_g is presumably stabilised. Accordingly five out of the six 3d electrons

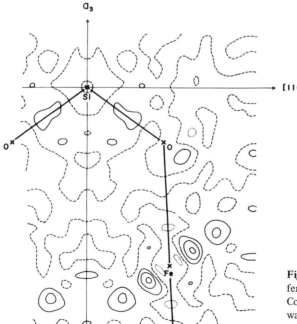

Fig. 5.23. A section of the difference Fourier map of γ-Fe_2SiO_4. Contours are drawn in a similar way to Fig. 5.21 (Marumo, Isobe and Akimoto, 1977)

occupy each of the five 3d orbitals, and the remaining one occupies the singlet level a_g in the ground state. The electron-density distribution in the half filled 3d shell has a spherical symmetry. The latter orbital a_g has a lobe elongated along the three-fold axis. The observed peaks of heights $1.2\ e\ Å^{-3}$ at $0.46\ Å$ from the Fe nucleus on the three-fold axis may originate in the electron in an a_g orbital.

X-ray evidence showing that these residual electron-density distribution are due to d electrons in non-bonding orbitals has been obtained by a study of electron-density distribution in rutile crystals. In rutile, a modification of TiO_2, the co-ordination around Ti is octahedral, while there is a triangular arrangement of the nearest neighbours around the oxide ions. Electron-density distribution in rutile crystals has been calculated based on a set of accurate intensity data (Shintani, Sato and Saito, 1975). The compound is non-stoichiometric and the exact formula is $TiO_{1.984}$. The crystal consists of Ti^{4+} and O^{2-} ions, thus the charge-density around titanium may be spherical, since Ti^{4+} possesses no d electrons. A section of the final difference synthesis through the titanium ion and perpendicular to the c axis is shown in Fig. 5.24. The resulting maps are featureless to within $0.3\ e\ Å^{-3}$. There are no meaningful peaks around the titanium atom, indicating that the charge-density around the Ti^{4+} is spherical.

The electron-density distribution in diaquaperoxo(2,6-pyridine-dicarboxylato)-titanium(IV), $[TiO_2(C_7H_3O_4N)(H_2O)_2] \cdot 2H_2O$, presents a striking contrast to that of rutile (Manohar and Schwarzenbach, 1974). Co-ordination around Ti is pentagonal bipyramidal. The titanium atom is almost neutral. In the final difference synthesis, there are maxima due to bonding electrons on every bond around the titanium atom in the molecule.

Asphericity of 3d charge-density in a tetrahedral environment has been examined for $CoAl_2O_4$ (Toriumi, Ozima, Akaogi and Saito, 1978). This has a normal spinel structure: the Co^{2+} ion is in the tetrahedral site and the Al^{3+} ion is in the octahedral site, however, the arrangement of Co and Al is somewhat disordered. The population of the Co^{2+} ion in the tetrahedral site is 84%. This means that 16% of the tetrahedral site is occupied by Al^{3+} ions. The calence-electron populations of the atoms were refined and the net charges of Co, Al and O atoms were estimated to be +1.5 (0.1),

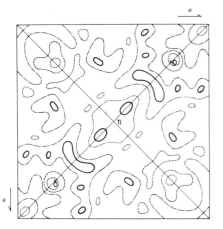

Fig. 5.24. A section of the difference Fourier synthesis through the titanium ion and perpendicular to the c axis. Contours are at intervals of $0.2\ e\ Å^{-3}$. Negative contours are broken lines, zero contours being chained (Shintani, Sato and Saito, 1975)

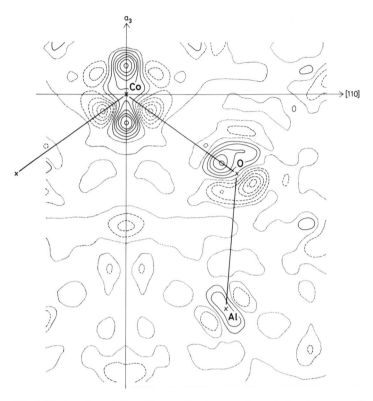

Fig. 5.25. A section of the difference Fourier map through the plane $y = x$. Contours are drawn at intervals of $0.2 \, e \, \text{Å}^{-3}$. Zero contours are in dotted lines, and negative contours are in broken lines (Toriumi, Ozima, Akaogi and Saito, 1978)

+2.8 (1) and −1.8 (1) e, respectively. Thus the compound is largely ionic. A section of the difference Fourier map through the plane $y = x$ is shown in Fig. 5.25, which corresponds to Fig. 5.21 for γ-Ni_2SiO_4. All the crystallographically independent atoms and chemical bonds are contained in this map. A remarkable feature is the distribution of residual electron-density around the Co^{2+} on the tetrahedral site: two positive peaks of $1.3 \, e \, \text{Å}^{-3}$ appear on the a_3 axis at 0.40 Å from the nucleus. In all, six peaks are located at the apices of an octahedron centred at the Co nucleus. On the other hand, four negative peaks appear on the three-fold axes through the Co and O atoms. Two out of the four negative peaks are at 0.40 Å from the Co toward to O atoms with peak heights of $-1.0 \, e \, \text{Å}^{-3}$. The remaining two with peak heights of -0.5 $e \, \text{Å}^{-3}$ are situated on the extension of the O–Co bond at about 0.5 Å from the Co nucleus (i.e. on the opposite side of the Co atom with respect to the O atom). These negative peaks are arranged at the eight corners of a cube centred at the Co nucleus. The arrangement of positive and negative peaks is contrary to that observed for the transition-metal ions in an octahedral environment. As described in Section 3 ii), the energy level of 3d electrons in a transition-metal atom placed in a tetrahedral environment splits into a lower doublet e and a higher triplet t_2 (Fig. 5.7). Then four out of seven d electrons of the Co^{2+} ion occupy the e orbital and the remaining three occupy the t_2 orbital in the ground state. In other words, two electrons are added in the e orbital to the half-closed shell ground state with spherical symmetry, $e^2t_2^3 + e^2$. Thus the expected deformation density around Co^{2+} may be the positive charge density in the direction of the e orbital and the negative charge density in the direction of the t_2 orbital. The observed deformation density shown in Fig. 5.25 exactly agrees with the expected one described above. Though not significant, residual density is observed around the O^{2-} ion on the Co–O bond. The positive peaks are directed twowards Co^{2+} and the negative peaks are at the rear of the O atom with respect to the Co atom. This indicates that the O^{2-} ion is polarised in the field of positive ions, being consistent with a slightly neutralised charge of the Co^{2+}. In other words, the Co–O bond is slightly covalent.

F. Charge Density Distribution in D_3 Complexes

Recently, the crystal structures of lel_3- and ob_3-isomers of the complexes $[M(chxn)_3]-(NO_3)_3 \cdot 3H_2O$ (M = Co, Rh) have been determined (Miyamae, Sato and Saito, 1977, 1979). The four crystals are isotypic and the complex ions are in a similar environment. Table 5.2 lists the effective charges on the central metal atoms together with those of $[Co(NH_3)_6]^{3+}$ and $[Co(CN)_6]^{3-}$. In the estimation of the effective charge, the radius of a sphere centred at the metal atom was taken as a distance between the metal nucleus and the peak due to bonding electrons. The central metal atom is largely neutralised owing to the donation of electrons from the ligating nitrogen atoms. Figure 5.26 illustrates schematically the change in the distribution of the residual charge density around the metal atom on descending the symmetry from O_h to D_3. Figure 5.26 (a) is the residual charge density (t_{2g}^6) with one of the three-fold axes being vertical. The six N atoms co-ordinate octahedrally. Figure 5.26 (b) illustrates what happens in the D_3 environment of a lel_3-isomer. In the complexes $[ML_6]$ the octahedron is twisted around the three-fold axis owing to the chelate ring formation.

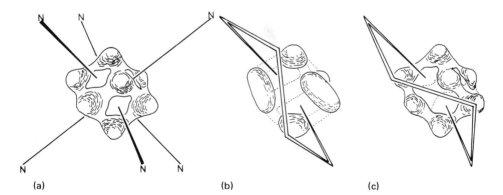

(a)   (b)   (c)

Fig. 5.26. Illustration of the change in residual electron density distribution around the metal atom **(a)** in O_h-, **(b)** $D_3(lel_3)$- and **(c)** $D_3(ob_3)$-environment. In (b) and (c) only one chelate ring is shown (Miyamae, Sato and Saito, 1979)

Table 5.2. Effective charge on the central metal atoms

Complex	Z	R[a]	$C(R)$	Effective charge	M-L bond length	Refs.
$[Co(NH_3)_6]^{3+}$	27	1.21 Å	26.2(0.1)	+0.8 e	1.972 Å	b
$[Co(CN)_6]^{3-}$	27	1.17	26.4(0.1)	+0.6	1.894	b
Λ-lel_3-$[Co(S,S-chxn)_3]^{3+}$	27	1.30	27.7(0.1)	−0.7[d]	1.972	b
Δ-ob_3-$[Co(S,S-chxn)_3]^{3+}$	27	1.30	27.7(0.1)	−0.7[d]	1.976	b
Δ-lel_3-$[Rh(R,R-chxn)_3]^{3+}$	45	1.52	44.6(0.2)	+0.4	2.071	c
Λ-ob_3-$[Rh(R,R-chxn)_3]^{3+}$	45	1.52	44.6(0.2)	+0.4	2.071	b

a   The distance between the central metal nucleus and the peak due to bonding electrons.
b   Miyamae, Sato and Saito, 1979.   c   Miyamae, Sato and Saito, 1977.
d   If R is taken to be 1.22 Å, the effective charge becomes +0.2.

The electrons in the non-bonding orbital have a tendency to avoid regions of high field due to ligand molecules. Thus the two lobes on the three-fold axis are more stable because no chelate ring spans across this direction. The six remaining peaks are fused to form three lobes in such a way that they keep away from the chelate ring plane. In the ob_3-isomer, Co−N bonds make larger angles with the three-fold axis than those in the lel_3-isomer and the C−C bond in the chelate ring is disposed nearly in the equatorial plane. Thus the residual charge-density near the equatorial plane differs from that in the lel_3-isomer. The change is schematically illustrated in Fig. 5.26 (c). The distribution avoids repulsion due to bonding electrons in the C−C bond. The symmetry of the distribution is no longer octahedral but D_3. Figure 5.27 (a) shows a section of the difference synthesis through a three-fold axis and a Co−N bond in the lel_3-$[Co(chxn)_3]^{3+}$ ion and in Fig. 5.27 (b) a section is shown through the cobalt atom perpendicular to the three-fold axis. Figure 5.27 (a) is analogous to Fig. 5.12. The sections of the three lobes can be clearly seen in Fig. 5.27 (b). The peaks on the triad axis are located at 0.52 Å from the cobalt nucleus, the height being 0.53 (0.12) e Å^{-3}. The peaks on the equatorial plane have heights of 0.51 (0.10) e Å^{-3} and 0.50 Å apart from the nucleus. The long axis of the lobe is inclined by about 11° with respect to the equatorial plane, thus the distribution is chiral. In the lel_3-Rh complex, 4d electrons spend most of their time outside the Ar core, on the other hand the geometry of the $[RhN_6]$ chromophore does not differ very much from that of $[CoN_6]$. Thus the electric field felt by the 4d electrons from the ligand is much stronger than by the 3d electrons. Accordingly the trend of non-bonding electron distribution is more marked. Figure 5.28 shows the difference synthesis of lel_3-$[Rh(chxn)_3]^{3+}$. There are five prominent peaks: two are on the three-fold axis at 0.61 Å from the Rh nucleus with heights of 0.29 e Å^{-3} and the remaining three are on the equatorial plane; they are located between the two adjacent chelate rings with peak heights of 0.44 e Å^{-3} at 0.53 Å from the Rh nucleus.

In the ob_3-isomer of cobalt, the distribution of non-bonding electrons shows the trend referred to earlier; however, the result is not very significant owing to poor intensity data. In the Rh ob_3-isomer, no significant peaks were observed in the equatorial plane, but the non-bonding electrons accumulate on the three-fold axis to

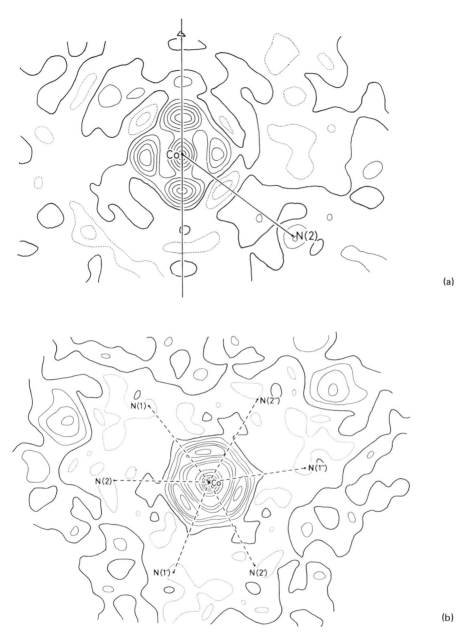

(a)

(b)

Fig. 5.27. Sections of the difference synthesis, lel_3-$[Co(chxn)_3]^{3+}$. (a) A section through a three-fold axis and a Co–N bond. (b) A section through the Co atom and perpendicular to the three-fold axis. Contours are drawn at intervals of $0.1\ e\ \text{Å}^{-3}$ (Miyamae, Sato and Saito, 1979)

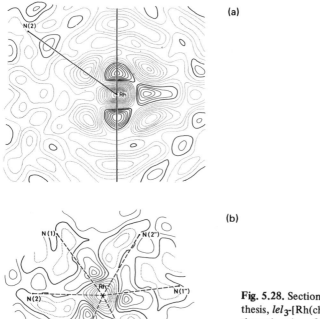

(a)

(b)

Fig. 5.28. Sections of the difference synthesis, lel_3-[Rh(chxn)$_3$]$^{3+}$. (a) A section through a three-fold axis and a Rh–N bond. (b) A section through the Rh atom and perpendicular to the triad axis. Contours are drawn at intervals of 0.05 e Å^{-3} in (a) and at 0.10 e Å^{-3} in (b) (Miyamae, Sato and Saito, 1977)

form large lobes: 0.48 e Å^{-3}, 0.64 Å from the Rh nucleus. The observed difference in residual charge distribution in the Co and Rh complexes seems to reflect the difference in the ligand field strength felt by the central metal atoms. The difference in residual electron density distribution in the lel_3- and ob_3-isomers indicates that the ground state wave function differs between the two isomers. In fact, the circular dichroism spectrum of the lel_3-isomer gives two peaks with opposite sign and different magnitudes, while that of the ob_3-isomer gives only a single peak in solution (see Table 6.3).

In the lel_3-[Rh(chxn)$_3$]$^{3+}$, a peak due to bonding electrons of height 0.15 e Å^{-3} in the Rh–N bond appears at 0.6 Å from the N atom (at 1.52 Å from Rh), slightly off the Rh–N bond. The peak-Rh-peak angle is 91°. This feature may indicate that the bonding orbital of the N atom is not directed exactly towards the Rh atom. Similar features were observed in the Co analogues but were not very significant. The same trend was observed in ob_3-[Co(chxn)$_3$]Cl$_3$ · H$_2$O (Kobayashi, Marumo and Saito, 1972).

5 Conclusions

To sum up, recent advances in techniques of the intensity measurements of diffracted
X-rays and the development of electronic computers have enabled us to determine
accurate charge-densities in transition metal complexes that have moderate complexity and are of sufficient chemical interest.

Firstly, the central metal atom is largely neutralised by donation of electrons
from the ligating atoms. Accordingly, Pauling's electroneutrality rule has indeed been
verified for transition-metal complexes. Secondly, charge-density of d-electrons in
non-bonding orbitals of a transition-metal atom in a complex is aspherical owing to
the ligand field. The distribution of non-bonding electrons has a tendency to
avoid regions of high field due to the ligands. The distribution can be reasonably
accounted for by ligand field theory. Thirdly, we can observe the distribution of
bonding electrons and the location of lone-pair electrons, and estimate the effective
charge on each atom with reasonable certainty. In some chelate complexes, the
"lone-pair" orbitals of the ligating atoms are not directed toward the central metal
atom. If the electron-density distribution and geometrical arrangement of the
atomic nuclei are all known, it is possible, at least in principle, to predict all the
physical and chemical properties of the complex on the basis of quantum mechanical calculations. In this respect, the accurate determination of the electron-density
distribution in transition-metal complexes will certainly play an important role for
the prediction as well as the rationalisation of the chemical and physical properties
of the complexes.

Chapter VI Circular Dichroism

1 General Introduction

It may be useful here to recollect briefly the physical optics of circular dichroism. The basic principles under discussion have already been covered by several recent books (Woldbye, 1965; Caldwell and Eyring, 1971; Hawkins, 1971; Ciarrdelli and Salvadori, 1973) and reviews (Mason, 1963; Schellman, 1975) to which the reader is referred for more details.

The vibration of a light wave is described by the form traced out by the electric vector of the light wave in a plane perpendicular to the direction of propagation. Figure 6.1 illustrates the types of polarisation with which we shall deal. The light wave is considered to propagate towards the direction of the observer's eye. Figure 6.1 (a) shows the vibration form of plane polarised light in which the electric field and the magnetic field vectors are confined to oscillate in two perpendicular planes. The plane on which the electric vector lies is usually referred to as plane of polarisation. We need not consider the magnetic vector, since this vector is perpendicular to

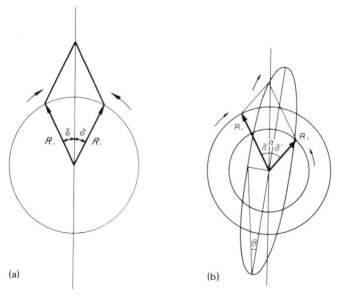

(a)   (b)

Fig. 6.1. Two vibration forms of a light wave, **(a)** plane polarised light and **(b)** elliptically polarised light

the electric vector. The figure also shows the resolution of plane (or linearly) polarised light into two interfering circularly polarised waves with opposite senses of rotation. The two component vectors rotate with the same angular velocity in opposite directions. Viewed in space the vector of circularly polarised light traces out a helix in which the pitch corresponds to the wavelength and the radius to the amplitude. Figure 6.1 (b) shows elliptically polarised light. The figure also shows that the elliptically polarised light can be resolved into two interfering circularly polarised lights with different amplitudes R_1 and R_r. The ellipticity is defined by the following equation:

$$\tan \theta = (R_r - R_l)/(R_r + R_l) \tag{6.1}$$

The refractive index of an isotropic medium is a measure of the velocity of light in the medium: the refractive index n of the medium with respect to vacuum is the ratio of the velocity of light in vacuum, c, to that in the medium, v, i.e.,

$$v = c/n \tag{6.2}$$

It should be noted that the frequency v of the light remains unchanged.

If the medium absorbs light of the wavelength under consideration, the intensity, I, decreases exponentially with the distance traversed. In the case of a solution, this relation can be expressed in terms of molar concentration, C, of the solute and decadic molar extinction coefficient, ϵ,

$$I = I_0 \times 10^{-\epsilon dC} \tag{6.3}$$

where I_0 is the incident beam intensity and d is the thickness of the layer.

Let us consider what happens when linearly polarised light travels through a layer of optically active medium. The left and right circularly polarised components pass through the medium with different velocity and are absorbed differently. These twin phenomena are called circular birefringence and circular dichroism respectively. As a result there will be a phase difference between the two components and the amplitudes are also different. When combined, they will give rise to elliptically polarised light, with the major axis of the ellipse inclined with respect to the original plane of polarisation. The situation can be formulated as follows: Let the refractive indices of left- and right-handed polarised light be n_l and n_r, respectively, and the molar extinction coefficients of the medium for both components be ϵ_l and ϵ_r, where we assume $n_l > n_r$ and $\epsilon_l > \epsilon_r$. Now Fig. 6.1 (a) is supposed to represent the cross section of the light wave at the moment when the light wave enters into the medium. The two component vectors of the left- and right-handed circularly polarised light are assumed to make an angle of δ with the original plane of polarisation. The amplitudes of the left- and right-circularly polarised component of the incident beam are both equal to $\sqrt{I_0}/2$. After traversing a distance, d, the intensities are reduced according to Eq. (6.3). The amplitudes of the left- and right-handed components, R_l and R_r, are given by

$$R_l = (\sqrt{I_0}/2) \, 10^{-(1/2)\epsilon_l dC} \tag{6.4}$$

and

$$R_r = (\sqrt{I_0}/2) \, 10^{-(1/2)\epsilon_r dC} \tag{6.5}$$

respectively. These components pass through the layer with different velocities:

$$v_l = c/n_l \quad \text{and} \quad v_r = c/n_r,$$

and thus the times required to travel the distance, d, are

$$d(n_l/c) \quad \text{and} \quad d(n_r/c),$$

respectively. The difference $d(n_l/c - n_r/c)$ corresponds to a phase difference of

$$2\pi v d(n_l - n_r)/c,$$

namely

$$2\pi d(n_l - n_r)/\lambda$$

since $c = v\lambda$.

The situation at the cross section separated by d is shown in Fig. 6.1(b). The figure shows the instant when the right-handed component makes an angle of δ with the original direction of polarisation. At this moment the left-handed component makes an angle of δ' with the original direction, which is given by

$$\delta' = \delta + 2\pi d(n_l - n_r)/\lambda \tag{6.6}$$

since the left-handed component lags behind the right-handed one by the amount given above.

Let the direction along which the resultant electric vector becomes maximum (direction of the major axis of the ellipse) be inclined at an angle α with the original plane of polarisation. We have

$$\delta + \alpha = \delta' - \alpha \tag{6.7}$$

since the two component vectors rotate with the same angular velocity.

From Eqs. (6.6) and (6.7), we obtain

$$\alpha = \pi d(n_l - n_r)/\lambda \tag{6.8}$$

And the ellipticity is given by

$$\tan\theta = (R_r - R_1)/(R_r + R_1) = [10^{-(1/2)\epsilon_r dC} - 10^{-(1/2)\epsilon_1 dC}]/$$
$$[10^{-(1/2)\epsilon_r dC} + 10^{-(1/2)\epsilon_1 dC}]$$

Multiplying the numerator and denominator by

$$10^{(\epsilon_r + \epsilon_1)dC/4}$$

we obtain

$$\tan\theta = [10^{(\epsilon_1 - \epsilon_r)dC/4} - 10^{-(\epsilon_1 - \epsilon_r)dC/4}]/[10^{(\epsilon_1 - \epsilon_r)dC/4} + 10^{-(\epsilon_1 - \epsilon_r)dC/4}]$$
$$= \tanh(\ln 10)dC(\epsilon_1 - \epsilon_r)/4$$

For small values of θ,

$$\theta \text{ (in radians)} \doteqdot \tan\theta \doteqdot \tanh\theta$$

Thus we have

$$\theta \doteqdot (\ln 10)/4\, dC(\epsilon_1 - \epsilon_r) = 0.576\, dC(\epsilon_1 - \epsilon_r) \tag{6.9}$$

To sum up, the electric vector traces out an ellipse in the same sense of rotation as that of the less absorbed circularly polarised component, i.e., $\epsilon_1 - \epsilon_r > 0$ corresponds to a clockwise rotation in Fig. 6.1(b). The major axis of the ellipse is rotated by an angle, $\pi d/\lambda(n_1 - n_r)$, i.e., $n_1 > n_r$ means dextro rotation.

The modern definition of specific rotatory power (or rotation), $[\alpha]_\lambda^t$, was originally introduced by Biot((1812, 1835). It is defined by

$$[\alpha]_\lambda^t = \alpha/l\rho \quad \text{for pure liquid}$$
$$= \alpha/lc \quad \text{for solution,}$$

where α is the observed rotation in degrees, l the path length in decimetres, ρ the density of the liquid and c is the concentration of the optically active solute in g/ml. The subscript and superscript stand for the wavelength of light used for the measurement and temperature, respectively. Molar rotation is obtained by multiplying the specific rotation by one hundredth of the molecular weight:

$$[\phi]_\lambda^t = M/100\, [\alpha]_\lambda^t$$

Ellipticity is a measure of circular dichroism. This is usually measured in *degrees* rather than in *radians*. The specific ellipticity, $[\psi]_\lambda$ is defined by analogy with the specific rotation, as $[\psi]_\lambda = \theta_\lambda/l\rho$ for pure liquid or θ_λ/lc for solution, where θ_λ is the measured ellipticity and the quantities l, ρ and c have the same significance as in specific rotation. Similarly, the molar ellipticity, $[\theta]_\lambda$, is given by

$$[\theta]_\lambda = [\psi]_\lambda M/100 \tag{6.10}$$

From Eq. (6.9)

$$[\theta]_\lambda = 2.303 \, \frac{4,500}{\pi} \, (\epsilon_l - \epsilon_r) = 3,300 \, (\epsilon_l - \epsilon_r) = 3,300 \, \Delta\epsilon \tag{6.11}$$

In practice, the molar coefficient of circular dichroic absorption can be measured conveniently by means of a dichrometer, thus, $\epsilon_l - \epsilon_r$ is by far the important quantity rather than the ellipticity itself. The circular dichroism, $\epsilon_l - \epsilon_r$ is exclusively used by chemists.

2 Interaction of Light with a Medium Containing Optically Inactive Molecules

A plane-polarised light wave moving in the positive z-direction has an oscillating electric field represented by the real part of

$$\mathbf{E} = \mathbf{E}_0 \exp i\omega(t - z/c) \tag{6.12}$$

and the magnetic field is given by

$$\mathbf{H} = \mathbf{H}_0 \exp i\omega(t - z/c) \tag{6.13}$$

where $|\mathbf{E}| = |\mathbf{H}|$ provided that $\mathbf{E}$ is expressed in electrostatic units and $\mathbf{H}$ is expressed in electromagnetic units. The oscillations are both in a plane perpendicular to the direction of propagation at right angles with each other. $\mathbf{E}_0$ is assumed to point in the positive x-direction, thus $\mathbf{H}_0$ points in the positive y-direction, if a right-handed co-ordinate system is used (Fig. 6.2). Let it fall perpendicularly on to a plane layer of thickness d

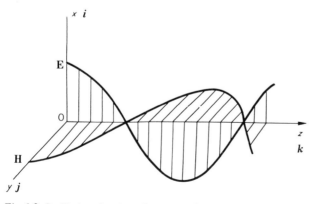

Fig. 6.2. Oscillating electric and magnetic field accompanied by a plane-polarised light moving in the positive z-direction

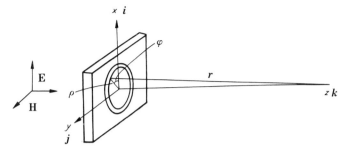

Fig. 6.3. Scattering of a plane-polarised light from a thin layer of molecules

containing N optically inactive molecules per unit volume, d being small compared to the wavelength λ of the incident light beam (Fig. 6.3). The electric field will induce a dipole moment

$$\mathbf{P} = \alpha \mathbf{E}_0 \exp i\omega(t - z/c) \tag{6.14)}$$

in each molecule, where α is the polarisability. This oscillating dipole will emit electric-magnetic radiation of the same frequency as the incident light. The magnitude of the electric field in the wavelet from the oscillating dipole observed at a distance $\mathbf{r}$ from the dipole is

$$|\mathbf{E}_s| = |\mathbf{E}_0| \, (\alpha\omega^2/rc^2) \sin \theta_s \exp i\omega(t - z/c)^{15)},$$

where θ_s is the angle between $\mathbf{E}_0$ and $\mathbf{r}$. The scattered spherical wavelet from each molecule will merge at some distance from the layer to form a plane wave front parallel to the wave front of the incident light. The incident wave front is parallel to the layer of molecules, so that at a given instant each molecule undergoes the same electric field and the induced dipole moment is the same for each molecule. The net electric field at a point located at a distance z from the plane can be calculated in the following way. First let us set up a polar co-ordinate system (ρ and φ) on the plane as shown in Fig. 6.3. From Fig. 6.3 symmetry consideration indicates that only the vertical component of $\mathbf{E}_s$ from all the molecules need be summed up, although $\mathbf{E}_s$ is perpendicular to $\mathbf{r}$. It can easily be shown that the magnitude of the vertical component is given by $|\mathbf{E}_s| \sin \theta_s$. Accordingly the net field is

---

15 This equation can be derived easily from the following fundamental equation relating an acceleration of a point charge and the electric field in the pulse emitted from it.

$$\mathbf{E} = (1/c^2 r^3) \, [\mathbf{r} \times \mathbf{r} \times \partial^2 \mathbf{P}/\partial t^2]$$

Since it takes time for an electromagnetic pulse to travel from the charge to an observer, the above expression gives the electric field due to an acceleration that occurred at a time r/c earlier. See for example, [Compton and Allison 1935].

$$E = E_0(\alpha\omega^2/c^2)Nd \exp(i\omega t) \int_0^{2\pi} \int_0^\alpha (1/r) \sin^2\theta \exp(-i\omega r/c)\rho d\rho d\varphi \qquad (6.15)$$

It can easily be shown that

$$\sin^2\theta = \sin^2\varphi + (z^2/r^2)\cos^2\varphi \qquad (6.16)$$

and

$$r^2 = \rho^2 + z^2 \qquad (6.17)$$

Inserting (6.16) and (6.17) into Eq. (6.15) and integrating over φ, we obtain

$$E_s = \pi E_0(\alpha\omega^2/c^2)Nd \exp(i\omega t) \int_0^\infty \rho/\sqrt{z^2 + \rho^2} \, [1 + z^2/(z^2 + \rho^2)]$$

$$\times \exp(-i\omega \sqrt{\rho^2 + z^2}/c)d\rho \qquad (6.18)$$

It is not difficult to show that the integral in Eq. (6.18) can be integrated by parts to give $-2ic/\omega \exp(-i\omega z/c)$ plus a term that varies as $c/\omega z^2$, which is negligible for sufficiently large $z \gg c/\omega$.

$$E_s = -2\pi i E_0(\alpha\omega/c)Nd \exp[i\omega(t - z/c)] \qquad (6.19)$$

It should be noted here that the factor $1/r$ in Eq. (6.15) has dropped in (6.19). This shows that the resultant is a plane wave, and the factor i indicates that the scattered wave is retarded in the phase angle by $\pi/2$ behind the incident wave.

Likewise the magnetic field also can induce a magnetic dipole, $\mathbf{M}$, in each molecule and this, in turn, emits electromagnetic radiation. The electric field strength observed at a distance $\mathbf{r}$ from the dipole is given by

$$\mathbf{E}_m = (1/c^2r^2)[\mathbf{r} \times \partial^2\mathbf{M}/\partial t^2] \qquad (6.20)$$

In deriving Eq. (6.19) it was assumed that the magnitude of electric field at the molecule in the layer is equal to that of the incident wave. Strictly speaking, this is incorrect. At optical frequencies, the field is increased by a factor of $(n^2 + 2)/3$, where n is the refractive index of the medium (Lorentz field). This correction can easily be made, when necessary, so that this fact was not considered.

---

16 Strictly speaking, this integral is ill-defined since the integrand oscillates with a finite amplitude even for large values of ρ. Such an integral can be made convergent by setting $\omega \to \omega - i\sigma$, where σ is an infinitesimal real positive parameter.

3 Interaction of Light with a Medium Containing Optically Active Molecules

If the molecule is optically active, it has an ability to give rise to an induced electric dipole moment by a changing magnetic field and an induced magnetic dipole moment by a changing electric field, in addition to an electric dipole moment induced according to Eq. (6.14). The induced electric and magnetic moments are thus given by

$$\mathbf{P} = \alpha\mathbf{E} - (\beta/c)\ \partial\mathbf{H}/\partial t \tag{6.21}$$

$$\mathbf{M} = (\gamma/c)\ \partial\mathbf{E}/\partial t \tag{6.22}$$

where β and γ are constants.

Putting Eqs. (6.12) and (6.13) into (6.21) and (6.22) respectively, we obtain

$$\mathbf{P} = \alpha E_0 \mathbf{i}\ \exp i\omega(t - z/c) - i(\omega/c)\beta E_0 \mathbf{j}\ \exp i\omega(t - z/c) \tag{6.23}$$

$$\mathbf{M} = i(\omega/c)\gamma E_0 \mathbf{i}\ \exp i\omega(t - z/c), \tag{6.24}$$

where $\mathbf{i}$ and $\mathbf{j}$ are unit vectors along $\mathbf{E}_0$ and $\mathbf{H}_0$ respectively.

These oscillating dipoles produce the scattered spherical wavelets and they combine to form a plane wave front at some distance from the surface which is large compared to the thickness d of the layer. The net wave motion can be calculated in the same way as described in the previous section on the basis of Eqs. (6.23) and (6.24). The resultant wave motion, when combined with the transmitted wave, is represented by

$$\mathbf{E}\ \exp i\omega(t - z/c) = \{[1 - 2\pi dN i(\omega/c)\alpha]E_0\mathbf{i} - 2\pi dN(\omega/c)^2$$
$$\times\ (\beta + \gamma)E_0\mathbf{j}\}\ \exp i\omega(t - z/c) \tag{6.25}$$

The second term in the square bracket can be ignored with respect to unity and we have

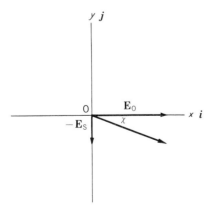

Fig. 6.4. Rotation of the plane of polarisation by an optically active medium

$$\mathbf{E} \exp i\omega(t - z/c) = (E_0\mathbf{i} - E_s\mathbf{j}) \exp i\omega(t - z/c) \tag{6.26}$$

where $E_s = 2\pi dN(\omega/c)^2 \, (\beta + \gamma)E_0$ \hfill (6.27)

When $\beta + \gamma$ is real, Eq. (6.27) represents a plane polarised wave propagating along the positive z-direction whose plane of polarisation differs from that of the incident wave. Referring to Fig. 6.28 the angle of rotation, χ, is given by

$$\tan \chi = \frac{2\pi\omega^2}{c^2} \, dN(\beta + \gamma) = (8\pi^3/\lambda^2) \, dN(\beta + \gamma) \tag{6.28}$$

When χ is small,

$$\chi = (2\pi\omega^2/c^2)dN(\beta + \gamma) \tag{6.29}$$

The angle χ is defined in such a way that if the plane of polarisation is rotated clockwise when one looks towards the source of the incident light, then χ is positive. This is in accordance with convention. If $\beta + \gamma$ is positive, it is dextro-rotatory and if it is minus, it is laevo-rotatory.

We shall try to generalise Eq. (6.29) for complex $\beta + \gamma$. In this case χ is also a complex and we put

$$\chi = \chi' + i\chi'' \tag{6.30}$$

$$\beta = \beta' + i\beta'' \tag{6.31}$$

$$\gamma = \gamma' + i\gamma'' \tag{6.32}$$

The electric field along the x and y axes can be written

$$E_x = E_0 = E \exp i\omega t \cos \chi \tag{6.33}$$

$$E_y = E_s = -E \exp i\omega t \sin \chi \tag{6.34}$$

If the co-ordinate axis is rotated clockwise by an angle χ', let the new co-ordinate axes be ξ and η.

$$E\xi = E_0 \cos \chi' - E_s \sin \chi' = E \exp i\omega t \cos (i\chi'') = E \exp i\omega t \cosh \chi'' \tag{6.35}$$

$$E\eta = E_0 \sin \chi' + E_s \cos \chi' = E \exp i\omega t \sin (-i\chi'') = -iE \exp i\omega t \sinh \chi'' \tag{6.36}$$

Since the electric field is given by the real parts of the above expressions, we have

$$E\xi = E \cos \omega t \cosh \chi'' \tag{6.37}$$

$$E\eta = E \sin \omega t \sinh \chi'' \tag{6.38}$$

It is evident that Eqs. (6.37) and (6.38) represent an ellipse with ξ and η as the principal axes. The elliticity is defined by

$$\tan \theta = - \sinh \chi''/\cosh \chi'' = - \tanh \chi'' \tag{6.39}$$

or $\quad \theta = - \chi'' \qquad$ for small χ''

The negative sign in the equation was taken in such a way that the electric vector represented by Eqs. (6.37) and (6.38) traces out the ellipse in the same sense of rotation as that of the circularly polarised component less absorbed in the medium i.e., $\epsilon_1 - \epsilon_r > 0$ corresponds to positive (clockwise rotation) within the ellipse (see Fig. 6.1). Thus Eq. (6.29) still holds for complex χ, β and γ, namely,

$$\chi = (2\pi\omega^2/c^2)dN(\beta + \gamma) \tag{6.40}$$

real part: $\chi' = \phi = (2\pi^2/c^2)dN(\beta' + \gamma') \tag{6.41}$

imaginary part: $\chi'' = - \theta = (2\pi\omega^2/c^2)dN(\beta'' + \gamma'') \tag{6.42}$

Thus the real part of $(\beta + \gamma)$ governs optical rotation and the imaginary part the ellipticity. Optical rotation and ellipticity (circular dichroism) depend on the wavelength of the incident light.

4 Quantum Theories

The quantum mechanical treatment of optical activity was initiated by Rosenfeld (1928). He used perturbation theory and showed that β and γ in Eqs. (6.21) and (6.22) are presented as follows:

$$\beta = \gamma = (c/3\pi h) \sum_a R_{0a}/(v_a^2 - v^2) \tag{6.43}$$

where h is Planck's constant, v is the frequency of the incident radiation and v_a is the frequency corresponding to transition from a ground state 0 to an excited state a. The summation is taken over all electronic transitions. R_{0a} is the rotatory strength of the transition and is given by

$$R_{0a} = \mathrm{Im} \langle 0|\mathbf{P}|a \rangle \cdot \langle a|\mathbf{M}|0 \rangle \tag{6.44}$$

where $\mathbf{P}$ and $\mathbf{M}$ are electric and magnetic dipole operators respectively. Thus R_{0a} is defined as the imaginary part of the scalar product of the electric and magnetic moments associated with the transition. The "bra" "ket" notation in Eq. (6.44) is defined as follows:

$$\langle 0|\mathbf{P}|a \rangle = \int \psi_0^* \mathbf{P}\psi_a d\tau \quad \text{etc.,}$$

where ψ_0 and ψ_a are the ground and excited state wave functions and an asterisk stands for the complex conjugate. This is called a matrix element of **P**.

It can be shown that

$$\sum_a R_{0a} = 0 \qquad (6.45)$$

From the definition of R_{0a} we have

$$\sum_a R_{0_a} = \mathrm{Im}[\sum_a \langle 0|\mathbf{P}|a\rangle \cdot \langle a|\mathbf{M}|0\rangle]$$

$$= \mathrm{Im}\,\langle 0|\mathbf{P}\cdot\mathbf{M}|0\rangle = 0$$

Since $\langle 0|\mathbf{P}\cdot\mathbf{M}|0\rangle$ is a diagonal element of a real observable, the imaginary part of $\langle 0|\mathbf{P}\cdot\mathbf{M}|0\rangle$ must be zero. This is called "sum rule" (Kuhn, 1929; Condon, 1937).

For an electronic transition to be optically active, its rotational strength, R_{0a}, must not be zero. This will impose certain restrictions on the symmetry of the molecule. It is not difficult to show that the rotational strength vanishes, if a molecule has a centre of symmetry or a mirror plane. The operators **P** and **M** have the form in Cartesian co-ordinate system,

$$\mathbf{P} = -e\,\sum_n (\mathbf{i}x_n + \mathbf{j}y_n + \mathbf{k}z_n) \qquad (6.46)$$

$$\mathbf{M} = (-eh/4\pi mc)\,i\,\sum_n [\mathbf{i}(z_n\,\partial/\partial y_n - y_n\,\partial/\partial z_n)$$

$$+ \mathbf{j}(x_n\,\partial/\partial z_n - z_n\,\partial/\partial x_n) + \mathbf{k}(y_n\,\partial/\partial x_n - x_n\,\partial/\partial y_n)] \qquad (6.47)$$

where **i**, **j** and **k** are unit vectors along the co-ordinate axes. The integrals such as $\langle 0|\mathbf{P}|a\rangle$ and $\langle a|\mathbf{M}|0\rangle$ may be evaluated in any co-ordinate system. In particular an integral may first be evaluated in an arbitrary system and then in the system obtained by applying one of the symmetry operations of the molecule: both must be identical.

If the molecule in question has a centre of symmetry, the states of the molecule can be classified as *odd* or *even* according to whether the wave function for a given non-degenerate state changes sign or retains the same sign when subjected to an inversion at the origin of co-ordinates:

$$x, y, z \;\rightarrow\; -x, -y, -z.$$

The operator **P** changes sign upon inversion, whereas the operator **M** does not change sign on inversion. Accordingly we have

$$\langle 0|\mathbf{P}|a\rangle \cdot \langle a|\mathbf{M}|0\rangle = -\,\langle 0|\mathbf{P}|a\rangle \cdot \langle a|\mathbf{M}|0\rangle$$

irrespective of the parity (even or odd nature) of the wave functions $|0\rangle$ and $|a\rangle$. Therefore

$\langle 0|\mathbf{P}|a\rangle \cdot \langle a|\mathbf{M}|0\rangle = 0$

and the optical rotatory power vanishes.

If the molecule has a mirror plane, R_{0a} vanishes, too. Let the mirror plane be the xy plane and again we classify the wave functions as odd or even with respect to reflection in this plane:

$x, y, z \rightarrow x, y, -z$

It is evident from the form of the operators given by (6.46) and (6.47) that the z-component of $\mathbf{P}$ is odd and the z component of $\mathbf{M}$ is even, whereas the x and y components of $\mathbf{P}$ are even and those of $\mathbf{M}$ are odd on reflection with respect to the xy plane. It can easily be seen that the scalar product $\langle 0|\mathbf{M}|a\rangle \cdot \langle a|\mathbf{M}|0\rangle$ is again identically zero leading to vanishing of rotational strength.

It is further possible to show that R_{0a} is identically zero, if the molecule has an improper axis of rotation in which a centre of inversion and a mirror plane are included. Molecules that lack the improper rotation axis all have the property of being non-superposable on their mirror images. This fact has traditionally served as a criterion for optical activity.

The Eq. (6.43) applies only outside the absorption regions, where $|\nu_a - \nu| \gg 0$. If $\nu = \nu_a$, the equation has no value. In classical theory, the absorption is treated by introducing the damping term in the equation of motion of electrons as in Eq. (2.20). The effect of damping appears in the denominator as an imaginary term, $i\gamma\omega$, in the steady state solution of Eq. (2.20).

$$P = -ex = (e^2/m) [1/(\omega_s^2 - \omega^2 + i\gamma\omega)] \tag{6.48}$$

In quantum theory the absorption is represented in terms of the lifetime of excited states and their spontaneous decay; however, the effect of absorption can be treated in exactly the same way as in classical theory. Thus the corresponding generalisation of equations to include the absorption region would be by a similar alteration of the denominator of Eq. (6.43).

$$\beta = \gamma = (c/3\pi h) \sum_a R_{0a}/(\nu_a^2 - \nu^2 + i\nu\Gamma_{0a}) \tag{6.49}$$

where Γ_{0a} is a positive constant and measures the strength of the damping and we assume $\Gamma_{0a} \ll \nu_a$.
Putting Eqs. (6.49) into Eq. (6.40), we have

$$\phi = (16\pi^2 N/3hc) \sum_a \nu^2 (\nu_{0a}^2 - \nu^2) R_{0a}/[(\nu_a^2 - \nu^2)^2 + \Gamma_{0a}^2 \nu^2] \tag{6.50}$$

$$\theta = (16\pi^2 N/3hc) \sum_a \nu^3 \Gamma_{0a} R_{0a}/[(\nu_a^2 - \nu^2)^2 + \Gamma_{0a}^2 \nu^2] \tag{6.51}$$

On rewriting these equations, we have

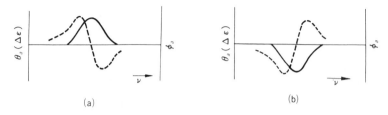

Fig. 6.5. The Cotton effect. A broken line represents optical rotation, ϕ_a and a full line ellipticity (circular dichroism). **(a)** $R_{0a} > 0$, **(b)** $R_{0a} < 0$

$$\phi = \sum_a \phi_a \qquad \phi_a = (16\pi^2 N/3hc)\nu^2(\nu_a^2 - \nu^2)R_{0a}/[(\nu_a^2 - \nu^2)^2 + \Gamma_{0a}^2\nu^2] \qquad (6.52)$$

$$\theta = \sum_a \theta_a \qquad \theta_a = (16\pi^2 N/3hc)\nu^3\Gamma_{0a}R_{0a}/[(\nu_a^2 - \nu^2)^2 + \Gamma_{0a}^2\nu^2] \qquad (6.53)$$

Figure 6.5 illustrates the changes of ϕ_a and θ_a within the region of absorption. If $R_{0a} > 0$, θ_a shows a positive peak at ν_a and ϕ_a is positive at the lower frequency side of the absorption region and changes sign on passing ν_a as shown in Fig. 6.5(a). If $R_{0a} < 0$, θ_a gives a negative peak at ν_a and ϕ_a behaves as illustrated in Fig. 6.5(b). It should be noted here that the sign of θ_a coincides with that of the rotational strength, R_{0a}.[17]

In 1896, Cotton discovered the effect which now bears his name. He observed anomalous rotatory dispersion and a circular dichroism band in the vicinity of an absorption band in solutions of potassium chromium(III) (or copper(II)) (+)-tartrate (Cotton, 1895, 1896). These two phenomena are called the Cotton effect. In this book, optical rotatory dispersion will not be considered further, since the circular dichroism is far more suitable for the study of co-ordination compounds. Unlike optical rotatory dispersion, the circular dichroism bands are restricted to the regions of absorption. This fact often facilitates the resolution of observed cd spectra into components for

---

17 Owing to the introduction of a term Γ_{0a}, θ_a (and also $\Delta\epsilon$) no longer causes catastrophe at $\nu = \nu_a$, and the behaviour of θ_a is close to what is actually observed. However, it should be noted that there are other strong reasons for the broadening of the absorption spectrum in addition to that described above. These are basically connected with the Frank-Condon principle. The transfer of d electrons from the ground state to higher d levels leads to excited states in which the equilibrium internuclear distances are greater than in the ground state. Thus the metal ion in its excited state interacts with its environment in a quantitatively different way from the same ion in its ground state. If the environment is somewhat variable, like a complex ion in a solution, the energy of the transition then depends on the momentary positions of neighbouring molecules and hence is itself slightly variable. This leads to a broadening of the absorption band. More generally, complex ions undergo thermal vibration, and if the optically excited molecule happens to be produced in a vibrationally excited condition, optical transition may occur at an instant when the internuclear distance is the same as that of the ground state. If so, many lines corresponding to different vibrationally excited states may appear so close that they cannot be distinguished from one another, resulting in a broad absorption band. The band shape depends upon the distribution of the energy levels due to vibronic coupling.

individual transitions. Moreover, the cd band shows a positive peak or a negative trough. This characteristic feature often enables us to resolve the two transitions as positive and negative cd peaks which are close in energy and this resolution is not possible in the ordinary absorption spectrum. Such resolution is essential for the assignment of absolute configuration, since the sign of the rotational strength is the crucial issue for the determination of absolute configuration.

The relation between the observed circular dichroism and the rotational strength can be derived as follows: consider an integral of θ_a/ν from $\nu = 0$ to $\nu = \infty$,

$$I = \int_0^\infty (\theta_a/\nu)d\nu$$

$$= (16\pi^2 N\Gamma_{0a}R_{0a}/3hc)\int_0^\infty \nu^2/[(\nu_a^2 - \nu^2)^2 + \Gamma_{0a}^2\nu^2]d\nu \tag{6.54}$$

As shown in Appendix VI-1, the definite integral in Eq. (6.54) has the value $\pi/(2\Gamma_{0a})$. Hence

$$R_{0a} = (3hc/8\pi^3 N)\int_0^\infty (\theta_a/\nu)d\nu \tag{6.55}$$

Combining Eqs. (6.55) and (6.9), we obtain

$$R_{oa} = [3hc10^3(\ln 10)/(32\pi^3 N_A)]\int_0^\infty (\Delta\epsilon/\nu)d\nu \tag{6.56}$$

where we put $d = 1$ cm, $C = 1$ mol/1,000 ml and $N = N_A/10^3$. This equation was first derived by Moffitt and Moscowitz (1959) by a different route. Thus the rotational strength may be obtained from the area of the corresponding circular dichroism band. For a Gaussian cd band with maximum at ν_0 and half width $\Delta\nu_{1/2}$, integration and insertion of the appropriate constants in (6.56) yield:

$$R_{0a} = 2.45 \times 10^{-39}(\epsilon_1 - \epsilon_r)_{max} \, \Delta\nu_{1/2}/\nu_0 \tag{6.57}$$

The sum rule for circular dichroism is then

$$\int[(\epsilon_1 - \epsilon_r)/\nu]d\nu = 0 \tag{6.58}$$

where the integration is carried out over the whole spectrum.

The rotational strength is a quantity analogous to the dipole strength, D_{0a}, which for a given transition represents the sum of the squares of electric dipole and magnetic dipole transition moments. The dipole strength is given, in c. g. s. units, from the area of the corresponding absorption band by

$$D_{0a} = [3hc\,10^3\,\ln 10/(8\pi^3 N_A)]\int (\epsilon/\nu)d\nu \tag{6.59}$$

where ϵ is the decadic molar extinction coefficient. The dipole strength is related to the classical oscillator strength, f_{0a}, by the equation,

$$f_{0a} = 8\pi^2 c v D_{0a}/(3he^2)$$ (6.60)

The oscillator strength gives the number of electrons promoted in the transition responsible for the absorption band.

5 Experimental Device; Dichrometer

The circular dichroism, $\Delta\epsilon$, is the difference in extinction coefficients for left and right circularly polarised light. Accordingly, the direct procedure to measure $\Delta\epsilon$ is to carry out the usual measurements of absorption by using circularly polarised light. An ingenious principle (Grosjean and Legrant, 1960) makes up the basis of commercial instruments (the Jouan, Cary and Jasco spectrophotometers). Figure 6.6 shows schematically a typical dichrometer. A monochromatic linearly polarised light beam is periodically transformed by means of a birefringent plate, M into a right and left circularly polarised light. The sample placed at S absorbs the two components differently. The rippling of the light intensity produces a d–c and an a–c signal at the photomultiplier and the periodic variation of the signal is proportional to $\Delta\epsilon$ which is to be measured. The transformation of the linearly polarised light into left and right circularly polarised light can be achieved by the electro-optic plate (Pockel's cell). The ammonium dihydrogen phosphate crystal, when exposed to longitudinal oscillating electric field, modulates the relative phase lag between the two circularly polarised components of the original linearly polarised light in such a way that their vibration form oscillates between the left and right circular polarisation. Alternatively, a photoelastic effect can be used to produce left and right circularly polarised light. An isotropic medium, when stressed periodically, exhibits a birefringence proportional to the stress. Such a plate can be used in place of Pockel's cell (Billardon and Badoz, 1966).

6 Optical Activity of Transition-Metal Complexes

After Cotton's and Werner's work (Cotton, 1895; Werner, 1911) the optical rotatory powers of transition-metal complexes were extensively studied in an attempt mainly to establish the relative configurations of the complexes. Notably Jaeger (1930) studied the optical rotatory dispersion and Mathieu (1946) the circular dichroism of Werner complexes. However, the theoretical configurations based on the proposed optical method (Jaeger, 1937) were not always consistent with the assignments of chirality based on solubility methods (Werner, 1912; Delépine, 1934). Until the advent of ligand field theory and the establishment of the absolute configurations of the complexes by X-ray methods, the electronic transitions responsible for the rotatory power

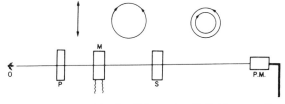

Fig. 6.6. The principle of a dichrometer, O: light source, P: polariser, M: modulator, S: sample, P. M.: photomultiplier. Upper diagrams show vibration forms of the light

of the metal complexes were not understood in any detail. In 1934 Kuhn and Bein developed a coupled oscillator model in order to correlate the absolute configuration to the sign and the form of the visible Cotton effects in chiral complexes (Kuhn and Bein, 1934a and b; Kuhn, 1938). An isotropic harmonic oscillator was assumed at the metal atom and three harmonic oscillators were placed along the edges of an octahedron spanned by chelate rings in a tris-chelated complex, such as $[Co(en)_3]^{3+}$. The charge displacements in each oscillator were correlated electrostatically with each other and with a linear charge displacement at the metal to generate an optical activity in the visible region. After a period of comparative neglect, ligand field theory, the basic idea of which had been originally developed by Bethe, Van Vleck and others during the period 1929 ~ 35, was applied to the interpretation of spectroscopic properties of transition-metal complexes in 1950's (Orgel, 1960 and Ballhausen, 1962). It was shown that the visible absorption band of the transition-metal complexes was due to the transition between d-levels split by the ligand field and the charge displacements at the metal atom were circular rather than linear. Saito and his collaborators gave a definitive answer to this problem by the first determination of the absolute configuration of $[Co(en)_3]^{3+}$ by the X-ray anomalous scattering methods. They showed that the result was enantiomeric to the model proposed by Kuhn and Bein. Moffitt (1956) introduced the first quantum mechanical theory of optical activity of chiral transition-metal complexes. He combined ligand field theory and "one-electron theory" of optical activity proposed by Condon, Altar and Eyring (1937) to derive expressions for the rotational strengths of the d–d transition. This work and those of other workers along this line provided the basis of recent spectroscopic and stereochemical applications.

In 1955, the tris(ethylenediamine)cobalt(III) isomer which is dextrorotatory at the sodium D line, $(+)_{589}[Co(en)_3]^{3+}$, was shown to have the Λ-absolute configuration. An absolute basis was thus provided for Mathieu's empirical relation. He had proposed that tris-chelated complexes with the same configuration as $(+)_{589}[Co(en)_3]^{3+}$ give a predominantly positive circular dichroism in the longest wavelength absorption band (1936). Since then, the number of complexes for which the absolute configuration has been studied by means of X-rays has grown at an increasing rate. At present, the absolute configurations of about 120 metal complexes have been established in this way.[18] They provide basic data for constructing a theoretical model for optical

18 A list of absolute configurations determined by the X-ray method up to the end of 1972 has been published (Saito, 1974).

activity and at the same time they give important criterion to evaluate many of the assignments of complex ion chirality on the basis of circular dichroism. At present it is possible to deduce the absolute configuration of the transition-metal complexes from the circular dichroism spectra with reasonable certainty.

7 Circular Dichroism Spectra of tris-Bidentate Complexes

A. Solution Circular Dichroism

Transition metal complexes of trigonal dihedral (D_3) symmetry have played a prominent role as model systems in both experimental and theoretical studies of optical activity, since they are stable and can be synthesized and resolved into optical isomers without difficulty. The high symmetry of the complexes makes detailed theoretical calculations more or less feasible. Tris(ethylenediamine)cobalt(III) ion is one of the most familiar and fundamental complex ions with D_3 symmetry and has been most extensively studied. Figure 6.7 shows the absorption and circular dichroism spectra of $(+)_{589}[Co(en)_3]^{3+}$. The values of rotational and dipole strengths are listed in Table 6.1.

When a cobalt(III) ion ($3d^6$) is placed in a ligand field of O_h symmetry, the five-fold degenerate orbitals are split into two groups, an upper doublet (e_g) and a lower triplet (t_{2g}) (octahedral splitting). On descending the symmetry to D_3, the lower triplet is further split into a stable singlet, a_1, and a higher doublet, e, (trigonal splitting). The absorption spectra of $[Co(en)_3]^{3+}$ consists of five bands: two bands of very weak intensity in near infrared or long wavelength visible region (A and B), two of moderate intensity in the visible and near ultraviolet region (the first and second absorption bands, I and II) and finally a very strong band in the ultraviolet region (CT). The excited states of the first four absorption bands are all $t_{2g}^5 e_g^1$ and these transitions are called d−d transitions, while the last is due to the ligand to metal charge-transfer. The corresponding energy terms are depicted in Fig. 6.8. and the assignments of the observed bands are indicated in Fig. 6.7. The absorption spectrum

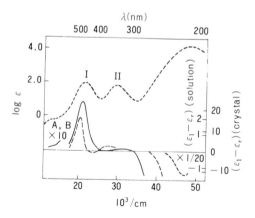

Fig. 6.7. The absorption (− − − −) and the circular dichroism spectrum (———) of $(+)_{589}[Co(en)_3]^{3+}$ in aqueous solution and the circular dichroism of the single-crystal $(+)_{589}[Co(en)_3]_2Cl_6 \cdot NaCl \cdot 6H_2O$ (−−−−) for light propagated along the c axis (optic axis)

Table 6.1. Electronic spectra and circular dichroism of $(+)_{589}[Co(en)_3]^{3+}$

		Electronic spectra			Circular dichroism		
		ν_{max} (10^3 cm^{-1})	ϵ	D^c	ν_{max} (10^3 cm^{-1})	$\epsilon_l - \epsilon_r$	R^c
Aqueous solution	A	13.7	0.35	4	13.7	+ 0.008	—
	I	21.3	84	1,200	20.3	+ 1.89	+ 4.2
					23.4	− 0.166	− 0.24
	II	29.4	74	950	28.5	+ 0.250	+ 0.48
	C.T.[b]	48.1	1.5×10^4	3×10^5	47.2	− 31	− 67
Single crystal[a]	A				14.0	+ 0.0025	+ 0.068 } at 80 K
	B				18.0	+ 0.15	+ 0.32
	I	21.4	95	1,500	21.1	+ 23.3	+ 79 } Room temp.
	II	29.4	110	1,500	29.0	+ 0.9	+ 2

a $(+)_{589}[Co(en)_3]_2Cl_6 \cdot NaCl \cdot 6H_2O$ (McCaffery and Mason, 1963; Mason and Peart, 1977) light beam parallel to the optic axis.

b Ligand to metal charge transfer band.

c $\times 10^{-40}$ c.g.s.

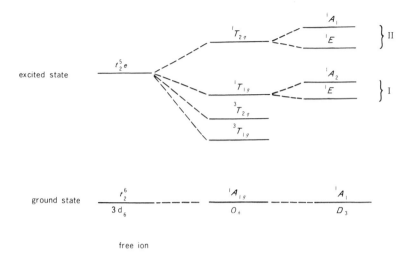

Fig. 6.8. The splitting of d orbital energy terms in O_h and D_3 environments for $3d^6$ configuration

of $[Co(en)_3]^{3+}$ of D_3 symmetry shows a great resemblance to that of octahedral $[Co(NH_3)_6]^{3+}$, since the visible absorption spectrum is largely determined by the nature of the ligating atoms, in both cases the chromophore being $[CoN_6]$. Actually the first absorption band consists of the two components, $^1A_1 \rightarrow {}^1E$ and $^1A_1 \rightarrow {}^1A_2$, however, these two transitions cannot be recognized as two separate bands in the first absorption region. The first absorption bands exhibit the largest rotatory power and two circular dichroism bands exist with opposite signs and of different magnitude in this region, reflecting the trigonal splitting. For a transition to be optically active, the associated magnetic moment must have a non-zero component along the direction of the associated electric dipole moment, as indicated by Eq. (6.44). The magnetic dipole selection rules for D_3 symmetry shown in Table 6.2. predict the occurrence of two cd components: $^1A_1 \rightarrow {}^1E$ and $^1A_1 \rightarrow {}^1A_2$ in band I and only one component $^1A_1 \rightarrow {}^1E$ in band II: another component $^1A_1 \rightarrow {}^1A_1$ is magnetic dipole forbidden. The observed circular dichroism spectra shown in Fig. 6.7 confirm all these predictions. The A_2 and E components are polarised respectively parallel and perpendicular to the three-fold axis of the complex ion. McCaffery and Mason (1963) measured the single crystal circular dichroism spectrum of $(+)_{589}[Co(en)_3]_2Cl_6 \cdot NaCl \cdot 6H_2O$ with light propagating parallel to the optic axis, in which all the complex ions are arranged with the three-fold axis parallel to it. In this condition only the E component is excited. As shown in Fig. 6.7, this crystal measurement showed that the intrinsic rotational strength of the $^1A_1 \rightarrow {}^1E$ transition is positive and substantially larger than that of solution circular dichroism. Accordingly the intrinsic rotational strength of $^1A_1 \rightarrow {}^1A_2$ must be negative and almost as large as that of the E component. Thus what is observed in a solution circular dichroism spectrum is a result of large cancellation due to the overlapping of the two rotatory strengths with opposite signs when the complex ions are randomly oriented

Table 6.2. Magnetic dipole selection rules for O_h and D_3 symmetries

	O_h				
	A_{1g}	A_{2g}	E_g	T_{1g}	T_{2g}
A_{1g}	x	x	x	o	x
A_{2g}	x	x	x	x	o
E_g	x	x	x	o	o
T_{1g}	o	x	o	o	o
T_{2g}	x	o	o	o	o

	D_3		
	A_1	A_2	E
A_1	x	∥	⊥
A_2	∥	x	⊥
E	⊥	⊥	∥, ⊥

o: allowed, x: forbidden; ∥: parallel to the C_3 axis;
⊥: perpendicular to the C_3 axis.

in solution. In fact, the band origins of the E and A_2 components lie close together
(Dingle, 1967) and the appearance of separate circular dichroism bands in the solution
spectrum originates in the different distribution of rotational strengths over the E and
A_2 vibronic progressions (Richardson, Caliga, Hilmes and Jenkins, 1975). The separa-
tion of the crystal circular dichroism spectra into E and A_2 components will be described in the next section. Table 6.3 lists the circular dichroism spectra of some tris-
bidentate complexes with five-membered chelate rings in the first absorption region (I).
With only one exception, $(+)_{589}[Co(cptn)_3]^{3+}$, those complexes which show prominent positive circular dichroism in the first absorption region possess Λ absolute configuration and if it is negative the absolute configuration is Δ. The empirical rule can
be slightly modified to include the case of $(+)_{589}[Co(cptn)_3]^{3+}$ as follows: those
complexes which show a positive circular dichroism band in the longer wavelength
side of the first absorption region have Λ absolute configuration.

The empirical rule mentioned above generally holds for most tris-chelated d^6
or d^3 complexes with five-membered chelate rings, even if the complex contains
ligating atoms other than nitrogen. Table 6.4 lists the circular dichroism spectra
of tris-chelated complexes containing six- or seven-membered chelate rings. The
absolute configuration of all the complexes listed in the table has been established
by means of X-rays. As can be seen from Tables 6.3 and 6.4, a series of complexes,
Λ-$[Co\{NH_2(CH_2)_nNH_2\}_3]^{3+}$ (n = 2, 3, 4), has a common feature in that they all
exhibit a positive circular dichroism band at the longer wavelength side of the first
absorption region, although the magnitudes differ considerably, and they all show
a negative circular dichroism band at the longest wavelength region of the charge
transfer absorption (not shown in Table 6.4).

Table 6.3. Circular dichroism spectra of tris-bidentate complexes of Co(III) and Cr(III) with five-membered chelate rings[a]

Complex	Chromo-phore	CD $\tilde{\nu}$ $10^3\,\mathrm{cm}^{-1}$	$\Delta\epsilon$	Absolute configuration	Refs.
$(+)_{589}[\mathrm{Co(en)_3}]^{3+}$	$[\mathrm{CoN_6}]$	20.28 23.31	+ 2.18 − 0.20	$\Lambda\,(\delta\delta\delta)\,lel_3$	d
$(+)_{589}[\mathrm{Co}(S\text{-pn})_3]^{3+}$	$[\mathrm{CoN_6}]$	20.28 22.78	+ 1.95 − 0.58	$\Lambda\,(\delta\delta\delta)\,lel_3$	e
$(+)_{589}[\mathrm{Co}(R\text{-pn})_3]^{3+}$	$[\mathrm{CoN_6}]$	21.0	+ 2.47	$\Lambda\,(\lambda\lambda\lambda)\,ob_3$	f
$(-)_{589}[\mathrm{Co}(S,S\text{-chxn})_3]^{3+}$	$[\mathrm{CoN_6}]$	20.0 22.5	+ 2.28 − 0.69	$\Lambda\,(\delta\delta\delta)\,lel_3$	g
$(+)_{589}[\mathrm{Co}(R,R\text{-chxn})_3]^{3+}$	$[\mathrm{CoN_6}]$	20.8	+ 3.9	$\Lambda\,(\lambda\lambda\lambda)\,ob_3$	g
$(+)_{589}[\mathrm{Co}(S,S\text{-cptn})_3]^{3+}$	$[\mathrm{CoN_6}]$	18.9 21.1	+ 0.59 − 1.91	$\Lambda\,(\delta\delta\delta)\,lel_3$	h
$(-)_{589}[\mathrm{Co(sar)(en)_2}]^{2+}$	$[\mathrm{CoN_5O}]$	19.4	− 1.8[b]	$\Delta\,(\lambda_{\mathrm{sar}}\delta_{\mathrm{en}}\lambda_{\mathrm{en}})$	i
$(+)_{495}[\mathrm{Co}(S\text{-glut})(\mathrm{en})_2]^{2+}$	$[\mathrm{CoN_5O}]$	19.6	+ 2.5[b]	$\Lambda\,(\delta\delta)$	j, n*
$(+)_{589}[\mathrm{Co}(S\text{-ala})_3]$	$[\mathrm{CoN_3O_3}]$	18.5 21.0	+ 1.3 − 0.2	Λ	k, o*
$(-)_{589}[\mathrm{Co(ox)_3}]^{3-}$	$[\mathrm{CoO_6}]$	16.2	+ 3.3	Λ	l, p*
$(+)_{546}[\mathrm{Co(thiox)_3}]^{3-}$	$[\mathrm{CoS_6}]$	15.8	− 0.2	Λ	m, q*
$(+)_{589}[\mathrm{Cr(ox)_3}]^{3-}$	$[\mathrm{CrO_6}]$	15.9 18.9	− 0.6 + 2.8	Λ	l, r*
$(+)_{589}[\mathrm{Cr(mal)_3}]^{3-\ \text{c}}$	$[\mathrm{CrO_6}]$	16.1 18.0	− 0.07 + 0.20	Λ	l

a   All the absolute configurations have been established by the X-ray method.
b   Taken from the figure.
c   Six-membered chelate rings.
d   Sato, Saito, Fujita and Ogino, 1974.
e   McCaffery, Mason and Ballard, 1965.
f   Douglas, 1965.
g   Piper and Karipides, 1964.
h   Ito, Marumo and Saito, 1971.
i   Buckingham, Mason, Sargeson, and Turnbull, 1966.
j   Dunlop, Gillard, Payne and Robertson, 1966.
k   Denning and Piper, 1966.
l   McCaffery and Mason, 1963.
m   Hidaka and Douglas, 1964.
n   Gillard, Payne and Robertson, 1970.
o   Drew, Dunlop, Gillard and Royers, 1966.
p   Butler and Snow, 1971b.
q   Butler and Snow, 1972.
r   Butler and Snow, 1971a.
*   Among the two references in a row, the one with an asterisk concerns the absolute configuration not described in Chapter IV.

Table 6.4. Circular dichroism spectra of tris-diamine complexes of cobalt (III) with six- and seven-membered chelate rings[a]

Complex	CD		Absolute configuration	Refs.
	$\tilde{\nu}$ $10^3 cm^{-1}$	$\Delta\epsilon$		
$(-)_{589}[Co(tn)_3]^{3+}$	18.69	+ 0.08	Λ C_3-chair$_3$	b
	30.0	− 0.17		
$(-)_{546}[Co(R,S\text{-ptn})_3]^{3+}$	20.0	− 6.1	Δ C_3-chair$_3$	c
$(-)_{546}[Co(R,R\text{-ptn})_3]^{3+}$	19.6	− 6.2	Δ $(\lambda\lambda\lambda)$ lel_3	d
$(+)_{546}[Co(R,R\text{-ptn})_3]^{3+}$	20.9	+ 26.8	Λ $(\lambda\lambda\lambda)$ ob_3	d
$(+)_{589}[Co(tmd)_3]^{3+}$	18.2	− 0.17	Δ $(\lambda\lambda\lambda)$ lel_3	e
	20.5	+ 1.83		

a   All the absolute configurations have been established by the X-ray method.
b   Gollogly and Hawkins, 1968.
c   Mizukami, Ito, Fujita and Saito, 1972.
d   Mizukami, Ito, Fujita and Saito, 1970.
e   Sato, Saito, Fujita and Ogino, 1974.

B. Tests for the Theoretical Models

Three sources of dissymmetry exist in the structures of octahedral complexes:

(i) Configurational dissymmetry
This is the dissymmetry arising from the distribution of chelate rings around the central metal atom and distortions of the ligating atoms from a regular octahedral arrangement.
(ii) Conformational dissymmetry
This is the inherent dissymmetry possessed by each chelate ring by virtue of its adopting a chiral conformation.
(iii) Vicinal effect
This is the optical activity induced in the d−d transitions by the substituents attached to the asymmetric centres.

These effects are empirically separable by adding or subtracting the observed circular dichroism spectra of appropriate complexes, since the potential exerted by the atoms in the ligands is additive [See Eq. (6.70)]. Where the configurational and vicinal effects are opposed, the former appears to be generally dominant. Usually substitution on the chelate ring is accompanied by smaller changes in ring conformation and distortion of the ligating atoms, which may give rise to a change in circular dichroism spectra. Thus vicinal effects may be rather indirect.

Attempts have been made to test the theoretical models for optical activity on the basis of solution cd and the known geometry of the complexes. The geometry and absolute configuration of $[ML_6]$ chromophores are relevant to the evaluation

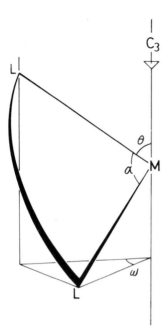

Fig. 6.9. Angles characterising a tris-bidentate complex

of the optical activity models of Piper and Karipides and of Richardson. According to theory (Karipides and Piper, 1964; Richardson 1971a, b, c, 1972a, b)[19], the trigonal splitting parameter K [= $2/3(\tilde{\nu}_E - \tilde{\nu}_{A_2})$] of the $^4A_{2g} \rightarrow {}^4T_{2g}$ (Cr^{III}) and $^1A_{1g} \rightarrow {}^1T_{1g}$ (Co^{III}) transitions may be controlled by whether an angle θ is greater or less than the octahedral value of 54.75°, where θ is an angle of inclination of the metal-ligand bond with respect to the three-fold axis of the complex ion as illustrated in Fig. 6.9. The angles characterising structures of some D_3 complexes of Cr^{III} and Co^{III} are listed in Table 6.5. It is known that the sign of the E component inverts from the circular dichroism spectrum of $\Lambda(+)_{589}$[Cr(ox)$_3$]$^{3-}$ to that of $\Lambda(+)_{589}$-[Cr(mal)$_3$]$^{3-}$ (McCaffery, Mason and Norman, 1964; McCaffery, Mason and Ballard, 1965). This is consistent with the observed change in θ. The reversal in component energies between $\Lambda(+)_{589}$[Co(thiox)$_3$]$^{3-}$ and $\Lambda(-)_{589}$[Co(ox)$_3$]$^{3-}$ is also in agreement with theoretical prediction (Butler and Snow, 1972). The situation is, however, not so straightforward. The [CoN$_6$] chromophore in $\Delta(-)_{589}$[Co(en)$_3$]$^{3+}$ is trigonally compressed and azimuthally contracted, whereas that in $\Delta(+)_{589}$-[Co(tmd)$_3$]$^{3+}$ is trigonally elongated and azimuthally contracted. In both these complexes, the longer wavelength band is known to be that of E symmetry (Sato and Saito, 1974). Thus the energy order of E and A$_2$ is not inverted whether the [CoN$_6$] chromophore is trigonally elongated or compressed. In fact, Strickland and Richardson (1973) calculated the rotatory strengths on a molecular-orbital model and showed that the d–d rotatory strength is very sensitive to details of the electronic structure within the chromophore.

---

19 See also Section D.

Table 6.5. Geometry of tris-bidentate complexes

	a	ω	θ	Refs.
$[Co(ox)_3]^{3+}$	84.3 (30)	54.0	56.4	a
$[Co(thiox)_3]^{3+}$	89.7 (2)	57.0	53.8	b
$[Cr(ox)_3]^{3-}$	82.4	50.3	56.3	c
$[Cr(mal)_3]^{3-}$	91.9 (6)	60.2	53.5	c
$lel_3\text{-}[Co(en)_3]^{3+}$	85.4 (3)	54.9	55.9	d
$lel_3\text{-}[Co(tmd)_3]^{3+}$	89.2 (2)	55.7	53.5	e

a   Butler and Snow, 1971b.
b   Butler and Snow, 1972.
c   Butler and Snow, 1971a.
d   Iwata, Nakatsu and Saito, 1969.
e   Sato and Saito, 1975b.

$[Co(TRI)_2]^{3+}$, $[Co(R\text{-}MeTACN)_2]^{3+}$ and $[Co(tame)_2]^{3+}$ have their non-ligating atoms above and below the trigonal planes formed by the ligating nitrogen atoms, while those of the tris-bidentate complexes such as $[Co(en)_3]^{3+}$, are between the opposed trigonal planes. The $[CoN_6]$ chromophore in $(+)_{546}[Co(TRI)_2]^{3+}$ possesses a small twist distortion around the three-fold axis. This is anti-clockwise as in the case of $\Lambda(+)_{589}[Co(en)_3]^{3+}$ which has a remarkably similar circular dichroism in the region of the first absorption band (Wing and Eiss, 1970). This similarity may constitute support for Piper's model. The $[CoN_6]$ chromophore in $(-)_{589}[Co(R\text{-}MeTACN)_2]^{3+}$ is elongated along and twisted around the triad axis ($\theta = 51.3°$, $\omega = 52.4°$). The twist direction is similar to that observed in $\Delta(-)_{589}[Co(en)_3]^{3+}$. The single-crystal circular dichroism shows a negative peak at 487 nm with light parallel to the optic axis, which is the trigonal axis of the complex ion, while that in an aqueous solution has a positive peak at about 477 nm. This observation indicates that the E band has a negative sign. The rotational strength of the A_2 transition seems to be greater than that of the E transition (Mikami, Konno, Kuroda and Saito, 1977). A similar study has been made with regard to $(+)_{589}[Co(tame)_2]^{3+}$ (Geue and Snow, 1977).

In fact, rotational strength may be classified as a second-order optical property and its extreme sensitivity to details of the electronic structure of the overall system requires very accurate theoretical calculations including all the atoms in the ligands and interactions between them.

C. Solid State Circular Dichroism of tris-Diamine Cobalt(III) Complexes

The physical properties of crystals, such as refractive index and extinction coefficient, are in general not the same for all crystal directions, owing to the fact that on passing through a crystal the sequence of atoms encountered depends upon the direction taken. It should be noted that in a crystal the refractive index and absorption coefficient depend not on the direction in which the electromagnetic waves are travelling but on the vibration direction of the electric vector. The phenomena are called double refraction and linear dichroism(pleochroism), respectively. Thus $\Delta\epsilon$ (= $\epsilon_l - \epsilon_r$) measured for a single-crystal plate cannot be directly compared with that measured for a solution, since ϵ depends on the instantaneous direction of vibration of the electric vector.

Crystals with cubic unit cells have the same atomic arrangements along all three axial directions. The optical properties are found to be the same not only along these three directions but also for all other directions, thus the cubic crystal is optically isotropic. Those crystals which belong to trigonal, hexagonal and tetragonal systems possess only one direction along which light travels with one refractive index(and also one absorption coefficient) independent from the vibration direction. This direction is called an optic axis and these crystals are called uniaxial. The optic axis is parallel to the four-fold axis in the case of tetragonal crystals and is parallel to the three-fold and six-fold axes for trigonal and hexagonal crystals respectively. A crystal plate cut perpendicular to the optic axis behaves isotropically for light waves propagating along the optic axis. Thus the measured circular dichroism spectra of such plates have the same significance as that obtained for a solution and the results may be compared without further correction. Crystals belonging to other crystal systems are biaxial, i.e., they possess two optic axes, but they have not been used for single crystal measurements before 1976 (Jensen). Mostly single-crystal measurements of circular dichroism are carried out for uniaxial crystals.

With the aid of the known crystal structures, it is possible to resolve the circular dichroism spectra of D_3 complexes in the first absorption region into the two components E and A_2 by combining single-crystal and the microcrystalline circular dichroism spectra of uniaxial crystals (Kuroda and Saito, 1976). Figure 6.10 illustrates the interaction of circularly polarised light with the complex ion in a uniaxial crystal. The light propagates along the optic axis OZ. The direction of the three-fold axis in the complex ion is indicated by a small closed triangle. The three-fold axis is inclined at an angle of α with respect to the optic axis. Let OY be a projection of the three-fold axis on a plane perpendicular to OZ. OX is chosen in such a way that OX, OY and OZ form a right-handed Cartesian co-ordinate system. When the circularly polarised light pro-

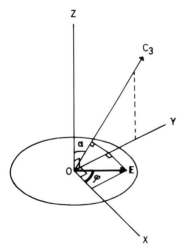

Fig. 6.10. Illustrating the interaction of circularly polarised light with the complex ion in a uniaxial crystal

pagates along OZ, its electric vector $\mathbf{E}$ rotates on the XY plane. The E component can be excited with the light whose electric vector is perpendicular to the three-fold axis of the complex ion, whereas the A_2 component can be excited when the electric vector is parallel to the three-fold axis. At a given instant, let the electric vector $\mathbf{E}$ make an angle of φ with OX. The intensities of the E and A_2 components excited at this instant will be proportional to

$$E^2 \cos^2 \alpha \sin^2 \varphi + E^2 \cos^2 \varphi \quad \text{for E} \tag{6.61}$$

and

$$E^2 \sin^2 \alpha \sin^2 \varphi \quad \text{for } A_2 \tag{6.62}$$

respectively. The average intensity over a period of rotation of the $\mathbf{E}$ vector will be obtained as follows:

$$\frac{1}{2} \pi \int_0^{2\pi} (E^2 \cos^2 \alpha \sin^2 \varphi + E^2 \cos^2 \varphi) d\varphi = E^2 (1 + \cos^2 \alpha)/2 \quad \text{for E} \tag{6.63}$$

and

$$\frac{1}{2} \pi \int_0^{2\pi} (E^2 \sin^2 \alpha \sin^2 \varphi) d\varphi = E^2 (\sin^2 \alpha)/2 \quad \text{for } A_2 \tag{6.64}$$

Since the E level is doubly degenerate, the difference between ϵ_l and ϵ_r of the single crystal, $\Delta\epsilon_s$, can be written as a sum of $\Delta\epsilon(E_X)$, $\Delta\epsilon(E_Y)$ and $\Delta\epsilon(A_2)$ as follows:

$$\Delta\epsilon_s = \frac{1}{4}(1 + \cos^2\alpha)/4 \Delta\epsilon(E_X) + \frac{1}{4}(1 + \cos^2\alpha)/4 \Delta\epsilon(E_Y) + \frac{1}{2}(\sin^2\alpha)/2 \Delta\epsilon(A_2) \quad (6.65)$$

On the other hand, in a microcrystalline state the three-fold axes of the complex ions are randomly oriented, hence $\Delta\epsilon$ in a microcrystalline state, $\Delta\epsilon_m$, can be written as

$$\Delta\epsilon_m = \frac{1}{3} \Delta\epsilon(E_X) + \frac{1}{3} \Delta\epsilon(E_Y) + \frac{1}{3} \Delta\epsilon(A_2) \tag{6.66}$$

where $1/3$ is a random orientation factor.
Since $\Delta\epsilon(E_X) = \Delta\epsilon(E_Y)$, Eqs. (6.65) and (6.66) become

$$\Delta\epsilon_s = \frac{1}{2}(1 + \cos^2\alpha)/2 \Delta\epsilon(E_X) + \frac{1}{2}(\sin^2\alpha)/2 \Delta\epsilon(A_2) \tag{6.67}$$

and

$$\Delta\epsilon_m = \frac{2}{3} \Delta\epsilon(E_X) + \frac{1}{3} \Delta\epsilon(A_2) \tag{6.68}$$

Now we can obtain the two simultaneous Eqs. (6.67) and (6.68) involving $\Delta\epsilon(E_X)$ and $\Delta\epsilon(A_2)$ as two unknowns. Using $\Delta\epsilon_s$ and $\Delta\epsilon_m$ measured at an arbitrary wavelength, $\Delta\epsilon(E_X)$ and $\Delta\epsilon(A_2)$ can be easily calculated by means of Eqs. (6.67) and (6.68). In this way the observed circular dichroism spectra can be resolved into E and A_2 com-

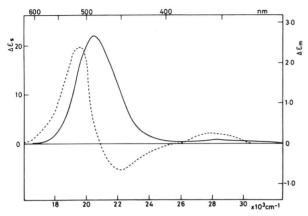

Fig. 6.11. The circular dichroism spectrum of a single-crystal (————) and that in a KBr matrix (————) of Λ-[Co(S,S-chxn)$_3$]Cl$_3$· 5H$_2$O

ponents. Figure 6.11 represents two circular dichroism spectra in the single-crystal and microcrystalline state of hexagonal Λ-[Co(S,S-chxn)$_3$]Cl$_3$ · 5H$_2$O. Figure 6.12 shows the result of resolution of the circular dichroism spectra shown in Fig. 6.11 into E and A$_2$ components.

By the procedure described above, the circular dichroism spectra of seven tris-(diamine)cobalt(III) complexes of known crystal structures were resolved into E and A$_2$ components. The results are summarised in Table 6.6, together with those obtained by other workers and are compared with the calculated values.

Jensen and Galsbøl (1977) directly measured the E and A$_2$ components by means of a phase modulation spectrophotometer (Hofrichter and Schellman, 1973). This

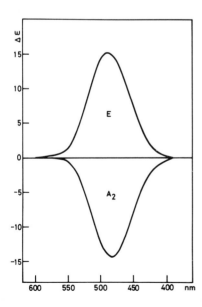

Fig. 6.12. The E and A$_2$ components obtained from the spectra shown in Fig. 6.11

Table 6.6. Rotatory strengths of tris(diamine) cobalt(III) complexes (10^{-40} c.g.s.). Crystal values of $R(E)$ and $R(A_2)$ are corrected by the fixed orientation factors of 2/3 and 1/3 respectively

| | Observed | | | | Calculated | | | | |
	$R(E)$	$R(A_2)$	$R(T_1)$	Refs.	$R(E)$	$R(A_2)$	$R(T_1)$	$R(T_1)_{expt}$	Refs.
Λ-2[Co(en)$_3$]Cl$_3 \cdot$ NaCl $\cdot$ 6H$_2$O	+62.9	−58.6	+4.3	a	+63.8	−59.8	+4.0	+4.4	e
	+52.6			b	+33.7	−30.4	+3.3		f
	+50.9			c					
	+43	−41		d					
Λ-[Co(en)$_3$]Br$_3 \cdot$ H$_2$O	+59.9	−55.7	+4.2	a	+63.8	−59.8	+4.0	+4.4	e
Λ-[Co(S-pn)$_3$]Br$_3$	+38.1	−36.6	+1.5	a	+65.1	−61.3	+3.8	+4.2	e
	+41.5			c					
Λ-[Co(S,S-chxn)$_3$]Cl$_3 \cdot$ 5H$_2$O	+56.5	−51.1	+5.4	a	+78.1	−74.0	+4.1	+3.9	e
Λ-[Co(S,S-cptn)$_3$]Cl$_3 \cdot$ 4H$_2$O^g	+57.3	−54.5	+2.8	a	+69.3	−65.4	+3.9	−4.3	e
Λ-[Co(S,S-ptn)$_3$]Cl$_3 \cdot$ 2H$_2$O^h	+12.5	−14.5	−2.0	a	+57.7	−55.3	+2.4	+1.9	e
Λ-[Co(tmd)$_3$]Br$_3$^{g}	+31.1	−38.7	−7.6	a	+93.0	−89.3	+3.7	−4.9	e

a Kuroda and Saito, 1976.
b McCaffery and Mason, 1963.
c Judkins and Royer, 1974.
d Jensen and Galsbøl, 1977.
e Mason and Seal, 1976.
f Evans, Schreiner and Hauser, 1974.
g These values are based on the cd spectra measured by Toftlund and Pedersen (1971).
h Experiments were carried out for the enantiomers.

spectrometer installs an optical modulator (see Fig. 6.6) which is a piece of fused silica. It is made to oscillate by mechanical coupling to a quartz crystal cut so that it has a resonance frequency of 50 kHz. The oscillation is selceted in such a way that it provides a periodically varying birefringence in the silica plate with an amplitude that is determined by the strength of the signal from the driving oscillator. After leaving the modulator the light passes through the sample and is received by a photomultiplier. The photocurrent is then sent through a lock-in-amplifier. By selecting the modulation appropriately a component proportional to $\Delta\epsilon(= \epsilon_l - \epsilon_r)$ can be detected by the electronic circuit (Jensen, 1976). In this way, the crystal circular dichroism spectra of Λ-(+)$_{589}$ tris(ethylenediamine)cobalt(III) ions diluted in a host crystal of racemic $2[Ir(en)_3]Cl_3 \cdot NaCl \cdot 6H_2O$ were measured. The crystal is hexagonal and the measurement was made with light propagating both parallel and perpendicular to the three-fold axis of the complex ion. The result is also included in Table 6.6.

As discerned from Table 6.6, the absolute values of the observed rotational strengths of the E and A_2 bands, separated by the method of Kuroda and Saito possess nearly the same magnitudes and opposite signs in conformity with theoretical prediction. The result was further supported by phase modulation spectrophotometric data (Jensen and Galsbøl, 1977). Thus it was established that what is observed in solution circular dichroism is the result of large cancellation of the two large components with different sign. $R(E)$ and $R(A_2)$ are unequal: $|R(E)| > |R(A_2)|$ for complexes with five-membered chelate rings, while $|R(E)| < |R(A_2)|$ for complexes with six- and seven-membered chelate rings. $R(T_1)$ is the sum of $R(E)$ and $R(A_2)$ and this value can be estimated by measuring the area under the observed circular dichroism curve. $R(T_1)$'s obtained from solid state circular dichroism spectra generally agree well in magnitude and sign with those obtained from solution circular dichroism spectra, except for Λ-[Co(S,S-cptn)$_3$]$^{3+}$, Λ-[Co(S,S-ptn)$_3$]$^{3+}$ and Λ-[Co(tmd)$_3$]$^{3+}$. For the latter three, some changes in conformation or in the mode of ion association may be expected in solution (Toftlund and Pedersen, 1971; Kuroda, Fujita and Saito, 1975).

D. Calculation of Rotatory Strengths
of tris-Bidentate Cobalt(III) Complexes

The rotational strengths of a complex ion can be calculated by means of Eq. (6.69):

$$R_{0a} = \text{Im} \langle 0|\mathbf{P}|a\rangle \cdot \langle a|\mathbf{M}|0\rangle \tag{6.69}$$

The wave function $|a\rangle$ is derived from a basis set $|M_k L_l)$ of simple products of metal ion and ligand functions, where M and L represent metal and ligand respectively. The perturbation is taken to be static potential between charge distribution:

$$V = \sum_{i(M)} \sum_{j(L)} q_i q_j / r_{ij} \tag{6.70}$$

The product function corrected to the first order of perturbation is

$$|M_a L_0\rangle = |M_a L_0\rangle + \sum_{kl} (E_a - E_k - E_l)^{-1} (M_k L_l|V|M_a L_0)|M_k L_l) \tag{6.71}$$

The d–d transitions are magnetic-dipole allowed but electric dipole forbidden. Thus the magnetic dipole moment of the transition $0 \rightarrow a$ has its zero-order value,

$$\langle M_a L_0|M|M_0 L_0\rangle = (M_a L_0|M|M_0 L_0) \tag{6.72}$$

The electric moment of the transition, however, is entirely borrowed with the following components expressed in the first order:

$$\langle M_0 L_0|P|M_a L_0\rangle = \sum_{k \neq 0} (-E_k)^{-1} (M_0 L_0|V|M_k L_0) (M_k|P|M_a) \tag{6.73a}$$

$$+ \sum_{k \neq 0} (E_a - E_k)^{-1} (M_k L_0|V|M_a L_0) (M_0|P|M_k) \tag{6.73b}$$

$$+ \sum_{l \neq 0} (-E_a - E_l)^{-1} (M_0 L_0|V|M_a L_l) (L_l|P|L_0) \tag{6.73c}$$

$$+ \sum_{l \neq 0} (E_a - E_l)^{-1} (M_0 L_l|V|M_a L_0) (L_0|P|L_l) \tag{6.73d}$$

where the ground state energy is taken to be zero. The first two terms, (6.73a) and (6.73b), on the right-hand side of the equation express the mixing of the electric-dipole forbidden and allowed transitions of the metal under the perturbations of the potential exerted by the ligands. These two terms form the basis of the ligand field one-electron model. It admits only static coupling between a chromophore electron localised on the metal atom and the static charge distribution located in the perturbing ligand in the ground state. The second two terms, (6.73c) and (6.73d) govern the dynamic coupling of the transition of the chromophore, $M_0 \rightarrow M_a$ with electric dipole transitions, $L_0 \rightarrow L_l$, induced in the ligand by the transition charge distribution.

In 1956, Moffitt first adopted the crystal field one-electron model and showed that the d–d transitions of metal complexes could become electric-dipole allowed under a static D_3 perturbing field, by mixing some p character into the d wave functions (Moffitt, 1956). An error in sign of the electric angular momentum operator, however, led to incorrect conclusions and Sugano (1960) subsequently demonstrated that Moffitt's model could not account for the net optical activity observed for the $^1A_1 \rightarrow {}^1T_1$ transitions in D_3 complexes of Co(III) in solution. In 1964 Piper and Karipides used a model in which d–d excitation obtained rotatory strength by borrowing intensity from charge-transfer bands (1964). In their treatment only the $[CoN_6]$ chromophore was considered. It was shown that the sign of rotational strength does not depend upon the absolute configuration of the chelate ring around the metal ion but is determined by the displacement of the ligating atoms from the apices of the regular octahedron. However, the atomic parameters determined by X-ray diffraction had to be violated in order to make good predictions. In the same year, Liehr used a different molecular orbital model (1964), in which the ligating nitrogen atoms were located on the regular octahedral axes. In this treatment the numerical values of rotational strength are determined by the absolute configuration of the complex ion and depend upon the magnitudes of the "angle of cant" between the axes of overlapping orbitals and Co—N axis, but are independent of the sign of this angle. When this model was proposed, it was very difficult to obtain information about the "angle of cant" of the overlapping orbitals. The model should, however, be reconsidered, since the direction of the overlapping orbitals can now be determined in the final difference synthesis (Chapter V).

Although these models could not fully account for the experimental results, the insight of these workers stimulated the work of a good many chemists. Other proposals were published based on the crystal field one-electron model without any marked advance (Hamer, 1962; Poulet, 1962; Bürer, 1963). Schäffer (1967, 1970) discussed the optical activity of D_3 complexes of cobalt(III) and chromium(III) in terms of the angular overlap model of bonding in co-ordination compounds first proposed by Yamatera (1957, 1958). This approach seems to show some promise, however, a detailed account of its application has not yet been given.

In all the models described above, only the first order contribution to the rotational strength was considered. But it appeared to be unrealistic in the crystal-field model, and the second order contributions to the net rotational strength were taken into account (Shinada, 1964; Caldwell, 1967; Richardson, 1971a, 1971b, 1972a, 1972b, 1971c). The theory could qualitatively account for the unequal magnitude and non-zero resultant rotational strength. Without a trigonal field splitting, however, the second order crystal-field contribution to the net rotational strength of a D_3 complex goes to zero. A trigonal field splitting in the complex ion, $[Co(en)_3]^{3+}$ is not detectable (Dingle, 1967). Strickland and Richardson (1973) refined the Piper and Karipides molecular orbital model by adding the ethylene portion of each ethylenediamine ligand in the form of a perturbing Coulombic field. They showed that neither the Piper molecular orbital model nor the Liehr model provides an adequate representation of the source of d-electron optical activity of D_3 complexes. Evans, Schreiner and Hauser (1974) adopted the static ligand field approach and calculated the optical activity of Λ-$[Co(en)_3]^{3+}$ on the basis of the molecular orbital model (1974). In the calculation, Co, C, N and H atoms were treated explicitly and allowed to interact whether they belonged to the same or different chelate rings and all the σ and π metal-ligand and ligand to ligand interactions were allowed. The calculated rotational strengths are compared with the experimental values in Table 6.6. The model accounts for the signs of the trigonal components, E and A_2, correctly and the magnitudes of the rotational strengths with considerable success.

The second model, based on dynamic coupling terms, (6.73c) and (6.73d), is basically identical with that developed by Tinoco (1962) and applied to organic systems by Hohn and Weigang (1968). It consists of independent metal ion and ligand groups. No overlapping is assumed between the electronic distributions of the individual groups. It is further assumed that the electronic properties of each group are approximately isolated from the rest of the system. Interaction between groups is then treated by perturbation theory. Obviously the crystal field one-electron model is subsumed in this more general theory. This model was adopted by Mason (1971) and Richardson (1971b, 1972a). These authors showed that the pairwise dynamic couplings between electrons in the ligand and the chromophoric d electron can provide significant contributions to the total d–d rotatory strength of chiral metal complexes. Mason and Seal (1976) calculated the optical activity of D_3 complexes of cobalt(III) on the basis of the dynamic coupling model. In their calculation, a Coulombic correlation was considered between the components of the electric hexadecapole moment of the $^1A_1 \rightarrow {}^1T_1$ d-electron transition of the Co(III) in the $[CoN_6]$ chromophore and a transient dipole induced in each ligand. Calculation of the rotatory strength based on this model accounted for the experimental result for $[Co(en)_3]^{3+}$ as shown in Table 6.6.

The theoretical calculations of Evans, Hauser and Schreiner as well as Mason and Seal correctly predict all the signs of $R(E)$ and $R(A_2)$. Moreover, theoretical values of $R(T_1)$ for the complexes with five-membered chelate rings agree well in sign and magnitudes with those observed in solid-state. Theoretical values of $R(E)$ and $R(A_2)$ for Λ-$[Co(en)_3]^{3+}$ agree satisfactorily with the observed values. For other five-membered chelate ring cases, $R(E)$ and $R(A_2)$ agree pretty well with the observation. For the six- and seven-membered ring cases the agreement is less satisfactory. No explanation can be offered at present.

Recently the single-crystal cd measurement at 80 K of Λ-$2[Co(en)_3]Cl_3 \cdot NaCl \cdot 6H_2O$ was extended to longer wavelength regions (Mason and Peart, 1977). The results are included in Table 6.1. The optical activity associated with spin-forbidden transitions to the $^3T_1$ and $^3T_2$ octahedral states was shown to possess distinctive features which are accounted for satisfactorily by spin-orbit coupling between these states and principally the corresponding $^1T_1$ state.

8 Circular Dichroism Spectra of cis-bis-Bidentate Complexes

The circular dichroism spectra of *cis*-bis-bidentate complexes of the general formulae, $[Co(a)_2(en)_2]^{n+}$ or $[Co(a)(b)(en)_2]^{n+}$, have also been extensively studied (McCaffery, Mason and Norman, 1965b). Generally they can be interpreted in relation to the parent complex, $[Co(en)_3]^{3+}$. On replacing one chelate ligand in the tris-diamine complex by a_2 the symmetry descends from D_3 to C_2. The first absorption band then gives rise to two transitions B_1 and B_2 and one A_2, of which A_2 and B_2 have trigonal E parentage. The splitting results in the shifting and splitting of absorption bands. The shift direction and mode of splitting depends upon the relative positions of the ligating atoms of a and N in the spectrochemical sereis. Table 6.7 lists the circular dichroism spectra of *cis*-bis-bidentate complexes of known absolute configuration determined by the X-ray method. The circular dichroism spectra show two peaks and the prominent peak can be interpreted as unresolved $A_2 + B_2$ of E trigonal parentage and its sign is diagnostic of the absolute configuration. This situation can be made clearer as follows: In Fig. 6.13 the wave numbers of circular dichroism peaks of three complexes are plotted of the type, Λ-*cis*-$[Co(a)_2(en)_2]^+$ against $\delta^1(a)$, where $\delta^1(a)$ is the difference between the wave numbers at the absorption band maxima of the two complexes $[Co(a)_6]$ and $[Co(NH_3)_6]^{3+}$. The two thick lines indi-

Table 6.7. Circular dichroism spectra of *cis*-bis-bidentate cobalt(III) complexes

Complex	Absolute configuration	CD $\tilde{\nu}$ $\times 10^3\,\mathrm{cm}^{-1}$	$\Delta\epsilon$	Refs.	
$(+)_{589}[CoCl_2(en)_2]^+$	Λ	16.3	-0.6	a	c**
		18.6	$+0.7$		
$(+)_{589}[Co(NO_2)_2(en)_2]^+$	Λ	21.7	$+1.4$	a	d
		25.0	-0.65		
$(+)_{589}[Co(CN)_2(en)_2]^+$	Λ	22.7	$+0.30$	a	e
		27.3	$+0.17$		
$(+)_{589}[Co(NO_2)_2(R\text{-}pn)_2]^+$	Δ	21.7	$-1.1*$	b	f
		24.5	$+0.6$		

* Taken from figure.
** References c ~ f concern the absolute configuration.
a McCaffery, Mason and Norman, 1965b.
b Barclay, Goldschmied, Stephenson and Sargeson, 1966.
c Matsumoto, Ooi and Kuroya, 1970.
d Matsumoto and Kuroya, 1972.
e Matsumoto, Ooi and Kuroya, 1972.
f Barclay, Goldschmied and Stephenson, 1970.

cate the expected positions for the two splitted components for $A_2 + B_2(E)$ and $B_1(A_2)$ (Yamatera, 1958). As can be seen from Fig. 6.13, all the prominent components lie on the line indicated as $(1/4)\delta^I(a)$ and the weaker ones are on the line indicated as $(1/2)\delta^I(a)$. Thus the relative peak positions of the $A_2 + B_2$ band and the B_1 band are inverted according to whether the ligand a comes prior to or behind en in the spectrochemical series. Accordingly the same empirical rule may be applied to these complexes as that which relates the absolute configuration of $[Co(en)_3]^{3+}$ to its circular dichroism spectrum: those complexes which show prominent positive circular dichroism in the first absorption region has the Λ absolute configuration.

For complexes of the type, *cis*-$[Co(a)(b)(en)_2]^{n+}$, the mode of splitting of the absorption band is similar and the assignment of the absolute configuration on the basis of circular dichroism spectra can be made in a similar way.

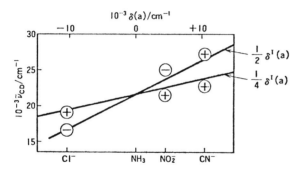

Fig. 6.13. Splitting of the circular dichroism spectra for Λ-*cis*-$[Co(a)_2 (en)_2]^+$ in the first absorption region

9 Sector Rules and the Circular Dichroism Spectra of Multidentate Complexes

A. Sector Rules

Various regional rules relating the stereochemistry to the optical activity of the transition-metal complexes have been proposed. Some of them are purely empirical and others possess a theoretical background. Notably two rules are known: one is Hawkins and Larsen's octant sign rule (1969) and the other is the "ring pairing method" of Legg and Douglas (1966). Both methods are essentially empirical and equivalent. The octant sign of a complex with chelate rings is derived by positioning the complex in a right-handed co-ordinate system so that the central metal atom is at the origin of co-ordinates and the donor atoms of a chelate ring lie in the xy plane and have the co-ordinates, $(+x, +y)$ and $(-x, +y)$. If the chelate ring in question is in an octant defined by $-x, -y, +z$, then the octant sign of this particular chelate ring is + (chelate ring 1). In the same way the chelate ring 2 can be assigned as +, i.e. $(+x, -y, -z)$. Thus the two rings that do not lie in the xy plane are completely contained in positive octants and the octant sign is positive. Each chelate ring has to be placed on the xy plane individually and the octant sign for each chelate ring is computed. Finally by adding up the signs the octant sign of the whole complex can be obtained. The octant sign tells us that if it is positive the circular dichroism spectra show a positive peak at the longer wavelength side in the first absorption region of Co(III) complexes with the chromophore [CoN_6]. The ring pairing method suggested by Legg and Douglas is as follows: For a given complex all possible combinations of two chelate rings are written down and the chirality (Λ or Δ) according to IUPAC convention (p. 9) of each set is determined. The net (or dominant) chirality should be governed by the chirality which occurs the greatest number of times. Figure 6.15 illustrates the application of this rule to $(+)_{589}[Co(penten)]^{3+}$. If the net chirality is Λ, the sign of the circular dichroism peak at the longer wavelength side in the first absorption region is positive and vice versa. In Legg and Douglas' original paper the use of Λ and Δ is opposite to that of the IUPAC nomenclature adopted here.

In addition to Co(III) and Cr(III) complexes, extensive studies have been made on Cu(II) and Ni(II) complexes with amino acids and polypeptides (Wellman, Mungall, Mecca and Hare, 1969; Wellman, Mecca, Mungall and Hare, 1968; Martin, Tsangaris

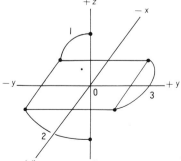

Fig. 6.14. Assignment of an octant sign for Λ-[Co(en)$_3$]$^{3+}$

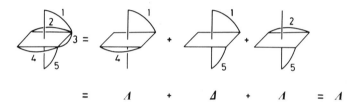

Fig. 6.15. Illustration of the procedure to obtain the net chirality of $(+)_{589}[Co(penten)]^{3+}$

and Chang, 1968; Wellman, Bogdansky, Piontek, Hare and Mathieson, 1969; Chang and Martin, 1969; Wilson and Martin, 1970; Tsangaris and Martin, 1970; Morris and Martin, 1971).

A theoretical analysis of these empirical rules can be accomplished by determining the symmetry-controlled aspects of the problem without performing the calculation of rotatory strength. In fact, a number of workers have succeeded in developing various sector rules. These rules are founded on the one electron theory of optical activity in dissymmetrically perturbed symmetric chromophores developed by Schellman (1966, 1968). The theory indicates that the induced rotatory power may be related to the substitution pattern by means of the symmetry properties of the unperturbed chromophore, independently of the detailed physical mechanism of the connection. In the case of centrosymmetric chromophores, the zero-order rotational strength may be non-zero and given by Eqs. (6.73a) and (6.73b). Schellman has shown that the potential V which mixes the transitions of the chromophore, transforms, or contains a component which transforms under the pseudoscalar representation of the point group to which the symmetrical chromophore belongs. If the rotatory strength (6.44) is non-zero, the matrix elements like $(M_k L_0 | V | M_a L_0)$ are totally symmetric, in other words, V is required to transform a product such as $M_k M_a$ or $(M_0 L_0 | P | M_k L_0)$ · $(M_a L_0 | M | M_0 L_0)$. This means that the function changes sign under all improper rotations but is invariant under all proper rotations of the point group. Since the potential function arises from outside the chromophore, it will not usually have the transformation properties of any representation of the symmetry groups of the chromophore. However, it is possible to express the potential function as a sum of such functions that have the transformation properties of the symmetry group of the chromophore (Wigner, 1959). There is an infinite number of those potential functions which transform as pseudoscalar under the symmetry operations of the point group. Among them only simpler potential functions may be useful for the sector rules. The functional forms of the simplest pseudoscalar potential are listed for most of the common point groups (Schellman, 1966, 1968).

Octant sign can be obtained by employing the co-ordinate function $z(x^2 - y^2)$ for each ligand atom (or substituent). The co-ordinate axes are right-handed and they are directed along the metal-ligand bonds. Along the scheme outlined above, Mason devised sector rules correlating the position of a substituent in the chromophores, $[CoA_5 B]$, trans-$[CoA_4 B_2]$ and $[CoA_6]$ (1970, 1971). Bosnich and Harrowfield (1972) presented a sector rule for the conformational isomers of octahedral complexes. The rule is based on a more intuitive argument utilizing experimental facts. Richardson

(1972b) derived the expression for rotational strength in which perturbation treatment was carried out to second order in both the wave functions and rotational strength. The second-order sector rules based on these expressions proved to be useful in relating the circular dichroism and absolute stereochemistry. These regional rules except the first two, octant sign rule and ring-pairing method, need, however, a more detailed geometry of the complex and hence are less practicable to predict the absolute configuration on the basis of circular dichroism spectra.

B. Circular Dichroism of Multidentate Complexes

Correlation of absolute configuration of multidentate complexes with their circular dichroism can be achieved by applying the octant rule or "ring-pairing method". Table 6.8 lists the circular dichroism spectra of the complexes containing multidentate ligands in the first absorption region. As shown in the Table, the two methods cover the main types of the chromophore and the sign of the Cotton effect due to a particular component descended from the octahedral T_{1g} transition in $[CoN_6]$, cis-$[CoN_4O_2]$ or cis-$[CoN_2O_4]$ chromophores can be correlated with the net chirality(or the octant sign). The relative magnitudes of the two circular dichroism peaks are affected by the ligands. For the complex ion, $(-)_{546}$-cis-β-$[Co(ox)(R,R,R,R$-3$''$, 2,3$''$-tet)]^+$, the prominent negative circular dichroism band appears to be shifted towards the shorter wavelength side in the first absorption region. It is to be noted here that the assignment of net chirality is impossible for those complexes in which three chelate rings join outside a face of an octahedron at non-coordinating atoms. In this case, the rings do not define the edges of an octahedron. Even if the number of skew chelate pairs could be counted, it is not possible to obtain net chirality. Thus the empirical rule mentioned above cannot be applied. Examples are the isomers of $[Co(S$-asp)_2]^-$ (Oonishi, Sato and Saito, 1975).

10 Conclusion

Both experimental and theoretical studies of transition-metal complexes of trigonal dihedral (D_3) symmetry have now enabled us to understand the origin of the optical activity of these complexes in reasonable detail. The theoretical models correctly account for the signs of the trigonal components, E and A_2, of the circular dichroism spectra of $[Co(en)_3]^{3+}$ and the magnitudes of the rotational strengths with considerable success. There, remains, however, much to be done before quantitative agreement can be obtained between theory and experiment for other complexes. On the other hand, the determination of the absolute configuration of a number of transition-metal complexes has established the empirical rules relating the absolute configuration and the circular dichroism spectra. We are now in a position to assign the absolute configuration of unknown complexes based on the circular dichroism spectra with reasonable certainty. In making asignments, reference complexes are needed

Table 6.8. Circular dichroism data of multidentate complexes of cobalt(III)

Complex	CD $\tilde{\nu}$ $10^3\,cm^{-1}$	$\Delta\epsilon$	Absolute configuration	Net chirality	Refs.
(+)$_{589}$-u-fac-[Co(dien)$_2$]$^{3+}$	19.9 22.5	+0.98 −0.84	ΛΔΛ	Λ	a
(+)$_{589}$-cis-β-[Co(NO$_2$)$_2$(R-2,2',2-tet)]$^+$	22.2 25.3	+2.50 −0.32	Λ	Λ	b, m**
(−)$_{589}$-cis-α-[Co(NO$_2$)$_2$ {3(S)8(S)-2',2,2'-tet}]$^+$	21.8 24.6	+1.60 −1.03	Λ	Λ	b
(−)$_{546}$-cis-β-[Co(NO$_2$)$_2$ {3(S)8(S)-2',2,2'-tet}]$^+$	21.9 25.0	−2.66 +0.43	Δ	Δ	b
(+)$_{589}$-cis-β[Co(CO$_3$) {3(S)8(S)-2'2,2'-tet}]$^+$	26.8 27.8	+2.74 −0.39	Λ	Λ	c,n**
(+)$_{546}$-cis-β-[Co(ox)(N,N'-Me$_2$-R,S-2,3",2-tet)]$^+$	17.5 20.0	−1.6* +0.8	ΔΔΔΔ	Δ	d***
(−)$_{546}$-cis-β-[Co(ox)(R,R-2,3",2-tet)]$^+$	18.0 21.0	+2.3* −1.3	ΛΛΛΔ	Λ	e***
(−)$_{546}$-cis-β-[Co(ox)(R,R,R,R-3",2,3"-tet)]$^+$	17.7 19.4	+0.64 −2.09	ΔΔΔΛ	Δ	f***
(−)$_{546}$-cis-β-[Co(edda)(R-pn)]$^-$	20.3	−1.7	ΔΔΔΛ	Δ	g
(+)$_{589}$[Co(limpen)]$^{3+}$	19.9 22.5	+1.68 −0.42	ΛΛΛΔ	Λ	h

$(+)_{589}[Co(penten)]^{3+}$	19.6 22.3	+3.61 −0.49	ΛΔΛ	Λ	i
$(−)_{589}[Co(R\text{-}mepenten)]^{3+}$	19.6 22.0	−3.31 +0.79	ΔΛΔ	Δ	j
$(+)_{546}[Co(edta)]^-$	17.0 19.4	−1.7 +0.9	ΔΛΔ	Δ	k
$(+)_{546}[Co(trdta)]^-$	16.7 18.9	−1.7* +2.5	ΔΛΔ	Δ	l

a   Kojima, Iwagaki, Yoshikawa and Fujita, 1977.
b   Saburi, Sawai and Yoshikawa, 1972.
c   Saburi, Sato and Yoshikawa, 1976.
d   Yano, Furuhashi and Yoshikawa, 1977.
e   Fujioka, Yano and Yoshikawa, 1975.
f   Yaba, Yano and Yoshikawa, 1976.
g   Halloran and Legg, 1974.
h   Yoshikawa and Yamasaki, 1973.
i   Schwarzenbach and Moser, 1953.
j   Gillard, 1964.
k   Ogino, Takahashi and Tanaka, 1970.
m   Tanaka, Marumo and Saito, 1973.
n   Toriumi and Saito, 1975.
*   Taken from figure.
** Absolute configuration not described in Chapter IV.
*** Circular dichroism and absolute configuration.

for which both crystal and molecular structure and the circular dichroism spectra are known in detail. Preferably these should have high conformational and configurational stability to minimize the possibility of structural change during phase changes from solid to solution.

The origin of the optical activity of dissymmetric transition metal complexes treated in this Chapter concerns the symmetric chromophore placed in a dissymmetric molecular field. There exists, however, another source of optical activity: an optically active complex is regarded as a dissymmetric ensemble of symmetric chromophores formed by co-ordination to a central metal atom. For instance, a tris-bidentate complex containing unsaturated ligands gives rise to optical activity by Coulombic coupling of the allowed $\pi \rightarrow \pi^*$ transition in the individual ligands. Owing to limited space, this topic has been completely excluded.

Appendix VI-1

Evaluation of the definite integral in Eq. (6.54).

Let I_1 be the definite integral in question

$$I_1 = \int_0^\infty v^2 / [(v_a^2 - v^2)^2 + \Gamma_{0a}^2 v^2] dv$$

Putting $v^2 = t$

$$I_1 = \frac{1}{2} \int_0^\infty \sqrt{t} / (t^2 - 2v_a^2 t + \Gamma_{0a}^2 t + v_a^4) dt$$

$$= \frac{1}{2} \int_0^\infty \sqrt{t} / \left[t^2 + 2 \left(\frac{\Gamma_{0a}^2}{2} - v_a^2 \right) t + v_a^4 \right] dt$$

By using the formula:

$$\int_0^\infty \frac{\sqrt{x}\,dx}{ax^2 + 2bx + c} = \frac{\pi}{\sqrt{2a}(\sqrt{ac} + b)}$$

cf. The Universal Encyclopedia of Mathematics, George Allen and Unwin Ltd. (1964), we obtain the desired result

$$I_1 = \pi / (2 \, \Gamma_{0a})$$

References

Allinger, N. L., Hirsch, J. A., Miller, M. A., Tyminski, I. J., VanCatledge, F. A.: J. Amer. Chem. Soc. *90*, 1199 (1968)

Allinger, N. L., Tribble, M. T., Miller, M. A., Werz, D. H.: J. Amer. Chem. Soc. *93*, 1637 (1971)

Anderson, B. F., Buckingham, D. A., Gainsford, G. J., Robertson, G. B., Sargeson, A. M.: Inorg. Chem. *14*, 1658 (1975)

Appleton, T. G., Hall, J. R.: Inorg. Chem. *10*, 1717 (1971)

Baggio, S., Becka, L. N.: Acta Crystallogr. *B 25*, 946 (1969)

Ballhausen, C. J.: Introduction to Ligand Field Theory New York: McGraw-Hill 1962

Barclay, G. A., Goldschmied, E., Stephenson, N. C., Sargeson, A. M.: Chem. Commun. *1966*, 540

Barclay, G. A., Goldschmied, E., Stephenson, N. C.: Acta Crystallogr. *B 26*, 1559 (1970)

Barlow, W.: Z. Krist, Mineral. *23*, 1 (1894); *25*, 86 (1895)

Beattie, J. K.: Acc. Chem. Res. *4*, 253 (1971)

Beddoe, P. G., Harding, M. J., Mason, S. F., Peart, B. J.: Chem. Commun. *1971*, 1283

Bensch, H., Witte, H., Wölfel, E.: Z. Physik. Chem. (Frankfurt) (N.S.) *1*, 256 (1954)

Bensch, H., Hermann, C., Peters, C.: Z. Physik. Chem. (Frankfurt) 4, 65 (1955)

Bijvoet, J. M., Peerdeman, A. F., van Bommel, A. J.: Nature, Lond. *168*, 271 (1951)

Billardon, M., Badoz, J.: Compt. rend. *B 262*, 1972 (1966)

Biot, J. B.: Mem. Inst. *1*, 1 (1812)

Biot, J. B.: Mem. Acad. Sci. *13*, 39 (1835)

Bixon, M., Lifson, S.: Tetrahedron *23*, 769 (1967)

Blount, J. F., Freeman, H. C., Sargeson, A. M., Turnbull, K. R.: Chem. Commun. *1967*, 324

Bosnich, B., MacHarrowfield, J. B.: J. Amer. Chem. Soc. *94*, 3425 (1972)

Boyd, R. H.: J. Chem. Phys. *49*, 2574 (1968)

Brill, R., Hermann, C., Peters, C.: Naturwissenschaften *32*, 33 (1944)

Brongersma, H. H., Mul, P. M.: Chem. Phys. Letters *19*, 217 (1973)

Brouty, C., Spinat, P., Whuler, A., Herpin, P.: Acta Crystallogr. *B 33*, 1913, 1970 (1977)

Buckingham, D. A., Mason, S. F., Sargeson, A. M., Turnbull, K. R.: Inorg. Chem. *5*, 1649 (1966)

Buckingham, D. A., Marzilli, L. G., Sargeson, A. M.: J. Amer. Chem. Soc. *89*, 825, 3428, 5133 (1967); Inorg. Chem. *6*, 1032 (1967)

Buckingham, D. A., Maxwell, I. E., Sargeson, A. M., Snow, M. R.: J. Amer. Chem. Soc. *92*, 3617 (1970)

Buckingham, D. A., Sargeson, A. M.: Top. Stereochem. *6*, 219 (1971)

Buckingham, D. A., Cresswell, P. J., Dellaca, R. J., Dwyer, M., Gainsford, G. J., Marzilli, L. G., Maxwell, I. E., Robinson, W. T., Sargeson, A. M., Turnbull, K. R.: J. Amer. Chem. Soc. *96*, 1713 (1974)

Bürer, Th.: Mol. Phys. *6*, 541 (1963)

Busch, D. H.: Helv. Chim. Acta, Fascilliculus Extraordinarius Alfred Werner, *1967*, 174

Busch, D. H., Farmery, K., Goedken, V., Katovic, V., Melnyk, A. C., Sperati, C. R., Tokel, N.: Adv. Chem. Ser., No. *100*, 44 (1971)

Butler, K. R., Snow, M. R.: Chem. Commun. *1971a*, 550; J. Chem. Soc. Dalton, *1976*, 251

Butler, K. R., Snow, M. R.: J. Chem. Soc., *A 1971b*, 565

Butler, K. R., Snow, M. R.: Inorg. Nucl. Chem. Lett. *8*, 541 (1972)

Caldwell, D. J.: J. Phys. Chem. *71*, 1907 (1967)
Caldwell, D. J., Eyring, H.: The Theory of Optical Activity. New York: Wiley-Interscience 1971
Chang, J. W., Martin, R. B.: J. Phys. Chem. *73*, 4277 (1969)
Ciadelli, F., Salvadori, P. (eds.): Fundamental Aspects and Recent Developments in Optical
    Rotatory Dispersion and Circular Dichroism. London: Heyden 1973
Clementi, E.: Tables of Atomic Functions IBM. J. Res. Dev., supplement *1965*, 9
Condon, E. U.: Rev. Mod. Phys. *9*, 432 (1937)
Condon, E. U., Altar, W., Eyring, H.: J. Chem. Phys. *5*, 753 (1937)
Coppens, P., Hamilton, W. C.: Acta Crystallogr. *A 26*, 71 (1970)
Coppens, P., Paulter, D., Griffin, J. F.: J. Amer. Chem. Soc. *93*, 1051 (1971)
Coppens, P.: Angew. Chem. Int. Ed. Engl. *16*, 32 (1977)
Corey, E. J., Bailar, J. C.: J. Amer. Chem. Soc. *81*, 2620 (1959)
Coster, D., Knol, K. S., Prins, J. A.: Z. Phys. *63*, 345 (1930)
Cotton, A.: Compt. rend. *120*, 989, 1044 (1895)
Cotton, A.: Ann. chim. et phys. *18*, 347 (1896)
Coulson, C. A.: Valence. Oxford: Clarendon Press 1961
Creaser, I. I., MacHarrowfield, J. B., Herlt, A. J., Sargeson, A. M., Springborg, J., Geue, R.J.,
    Snow, M. R.: J. Amer. Chem. Soc. *99*, 3181 (1977)
Crossing, P. F., Snow, M. R.: J. Chem. Soc. Dalton *1972*, 295
Curtis, N. F.: Coord. Chem. Rev. *3*, 3 (1968)

DeCoen, J. L., Elefante, G., Liquori, A. M., Damiani, A.: Nature, Lond. *216*, 910 (1967)
Delépine, M.: Bull. Soc. chim. France *1*, 1256 (1934)
Denning, R. G., Piper, T. S.: Inorg. Chem. *5*, 1056 (1966)
Dingle, R.: J. Chem. Phys. *46*, 1 (1967)
Douglas, B. E.: Inorg. Chem. *4*, 1813 (1965)
Drew, M. G. B., Dunlop, J. H., Gillard, R. D., Royers, D.: Chem. Commun. *1966*, 42
Duesler, E. N., Raymond, K. N.: Inorg. Chem. *10*, 1486 (1971)
Dunlop, J. H., Gillard, R. D., Payne, N. C., Robertson, G. B.: Chem. Commun. *1966*, 874
Dwyer, F. P., Garvan, F. L.: J. Amer. Chem. Soc. *81*, 1043 (1959)
Dwyer, F. P., MacDermott, T. E., Sargeson, A. M.: J. Amer. Chem. Soc. *85*, 2913 (1963)
Dwyer, F. P., Sargeson, A. M., James, L. B.: J. Amer. Chem. Soc. *86*, 590 (1964)
Dwyer, M., Searle, G. H.: Chem. Commun. *1972*, 726
Dwyer, M., Geue, R. J., Snow, M. R.: Inorg. Chem. *12*, 2057 (1973)

Evans, R. S., Schreiner, A. F., Hauser, P. J.: Inorg. Chem. *13*, 2185 (1974)

Fedrow, E.: Trans. Russian Min. Soc. *21*, 1; *25*, 1 (1885)
Flook, R. J., Freeman, H. C., Scudder, M. L.: Acta Crystallogr. *B 33*, 801 (1977)
Freeman, F. A., Liu, C. F.: Inorg. Chem. *7*, 764 (1968)
Freeman, H. C., Maxwell, I. E.: Inorg. Chem. *8*, 1293 (1969)
Freeman, H. C., Maxwell, I. E.: Inorg. Chem. *9*, 649 (1970)
Freeman, H. C., Marzilli, L. G., Maxwell, I. E.: Inorg. Chem. *9*, 2408 (1970)
Fujioka, A., Yano, S., Yoshikawa, S.: Inorg. Nucl. Chem. Letters *11*, 341 (1975)
Fujita, M., Yoshikawa, Y., Yamatera, H.: Chem. Letters *1976*, 959

Geue, R. J., Snow, M. R.: J. Chem. Soc. *A 1971*, 2981
Geue, R. J., Snow, M. R.: Inorg. Chem. *16*, 231 (1977)
Gillard, R. D.: Spectrochim. Acta *20*, 1431 (1964)
Gillard, R. D., Payne, N. C., Robertson, G. B.: J. Chem. Soc. *A 1970*, 2579
Gleicher, G. J., Schleyer, P. R.: J. Amer. Chem. Soc. *89*, 582 (1967)

Gollogly, J. R., Hawkins, C. J.: Chem. Commun. *1968,* 689
Gollogly, J. R., Hawkins, C. J.: Inorg. Chem. *8,* 1168 (1969)
Gollogly, J. R., Hawkins, C. J.: Inorg. Chem. *9,* 576 (1970)
Gollogly, J. R., Hawkins, C. J., Beattie, J. R.: Inorg. Chem. *10,* 317 (1971)
Gollogly, J. R., Hawkins, C. J.: Inorg. Chem. *11,* 156 (1972); see also Hawkins (1971)
Göttlicher, S., Wölfel, E.: Z. Elektrochem. *63,* 891 (1959)
Grosjean, M., Legrand, M.: Compt. rend. *251,* 2150 (1960)

Halloran, L. J., Legg, J. I.: Inorg. Chem. *13,* 2193 (1974)
Halloran, L. J., Caputo, R. E., Willet, R. D., Legg, J. I.: Inorg. Chem. *14,* 1762 (1975)
Hamer, N. K.: Mol. Phys. *5,* 339 (1962)
Harnung, S. E., Kallesbøe, S., Sargeson, A. M., Schäffer, C. E.: Acta Chem. Scand. *A 28,* 385 (1974)
Harnung, S. E., Sørensen, B., Creaser, L., Malgaard, H., Pfenninger, U., Schäffer, C. E.: Inorg. Chem. *15,* 2123 (1976)
Harris, H. A.: Ph. D. Thesis, Yale University 1966
Haupt, H. J., Huber, F., Preut, H.: Z. anorg. allg. Chem. *422,* 255 (1976)
Hawkins, C. J., Larsen, E.: Acta Chem. Scand. *19,* 185 (1969)
Hawkins, C. J.: Absolute Configuration of Metal Complexes. New York: Wiley, Interscience 1971
Heitler, H., London, F.: Z. Physik *44,* 455 (1927)
Helm, F. T., Watson, W. H., Radanovic, D. J., Douglas, B. E.: Inorg. Chem. *16,* 2351 (1977)
Hidaka, J., Douglas, B. E.: Inorg. Chem. *3,* 1724 (1964)
Hill, T. L.: J. Chem. Phys. *16,* 399 (1948)
Hirschfelder, J. O., Curtis, C. F., Bird, R. B.: The Molecular Theory of Gases and Liquids. New York: Wiley 1954
Hofrichter, H. J., Schellman, J. A.: Jerusalem Symposia on Quantum Chemistry V. Jerusalem: The Israel Academy of Sciences and Humanities 1973. See also, Jensen, H. P., Schellman, J. A., Troxell, T.: Appl. Spectroscopy *32,* 192 (1978)
Hohn, E. G., Weigang, O. E.: J. Chem. Phys. *48,* 1127 (1968)
Hönl, H.: Z. Phys. *84,* 1 (1933a)
Hönl, H.: Ann. Phys. (Leipzig) *18,* 625 (1933b)
House, D. A., Garner, C. S.: Inorg. Chem. *5,* 2097 (1966)

International Tables for X-ray Crystallography, Vol. I. Birmingham: Kynoch Press 1959
Ito, M., Marumo, F., Saito, Y.: Acta Crystallogr. *B 27,* 2187 (1971)
Ito, M., Marumo, F., Saito, Y.: Acta Crystallogr. *B 28,* 457 (1972a)
Ito, M., Marumo, F., Saito, Y.: Acta Crystallogr. *B 28,* 463 (1972b)
IUPAC: Inorg. Chem. *9,* 1 (1970)
IUPAC, International Union of Pure and Applied Chemistry: Nomenclature of Inorganic Chemistry, 2nd ed. London: Butterworths 1971
Iwasaki, H., Saito, Y.: Bull. Chem. Soc. Jpn. *39,* 92 (1966)
Iwasaki, H.: Acta Crystallogr. *A 30,* 173 (1974)
Iwata, M., Nakatsu, K., Saito, Y.: Acta Crystallogr. *B 25,* 2562 (1969)
Iwata, M., Saito, Y.: Acta Crystallogr. *B 29,* 822 (1973)
Iwata, M.: Acta Crystallogr. *B 33,* 59 (1977)

Jaeger, F. M., Blumendal, H. B.: Z. anorg. u. allgem. Chem. *175,* 161 (1928)
Jaeger, F. M.: Spacial Arrangements of Atomic Systems and Optical Activity, George Fisher Baker Lectures, Vol. 7, Cornell University, New York: McGraw-Hill 1930
Jaeger, F. M.: Proc. K. Ned. Akad. Wet. *40,* 108 (1937)
James, L. W., Antypas, G. A., Edgecombe, J., Moon, R. L., Bell, R. L.: J. Appl. Phys. *42,* 4976 (1971)
Jensen, H. P.: Acta Chem. Scand. *A 30,* 137 (1976)

Jensen, H. P., Galsbøl, F.: Inorg. Chem. *16*, 1294 (1977)
Johansen, H.: Acta Crystallogr. *A 32*, 353 (1976)
Judkins, R. R., Royer, D. J.: Inorg. Chem. *13*, 945 (1974)
Jurnak, F. A., Raymond, K. N.: Inorg. Chem. *11*, 3149 (1972)
Jurnak, F. A., Raymond, K. N.: Inorg. Chem. *13*, 2387 (1974)

Kaas, K., Sørensen, A. M.: Acta Crystallogr. *B 29*, 113 (1973)
Kalman, B. L., Richardson, J. W.: J. Chem. Phys. *55*, 4443 (1971)
Kamimura, H., Koide, S., Sugano, S., Tanabe, Y.: J. Phys. Soc. Jpn. *13*, 464 (1958)
Karipides, A., Piper, T. S.: J. Chem. Phys. *40*, 674 (1964)
Keene, F. R., Searle, G. H., Yoshikawa, Y., Imai, A., Yamasaki, K.: Chem. Commun. *1970*, 784
Kemp, J. D., Pitzer, K. S.: J. Amer. Chem. Soc. *59*, 276 (1937)
Kim, P. H.: J. Phys. Soc. Jpn. *15*, 445 (1960)
Kobayashi, A., Marumo, F., Saito, Y.: Acta Crystallogr. *B 28*, 2709 (1972a)
Kobayashi, A., Marumo, F., Saito, Y.: Acta Crystallogr. *B 28*, 3591 (1972b)
Kobayashi, M., Marumo, F., Saito, Y.: Acta Crystallogr. *B 28*, 470 (1972c)
Kobayashi, A., Marumo, F., Saito, Y.: Acta Crystallogr. *B 29*, 2443 (1973)
Kobayashi, A., Marumo, F., Saito, Y.: Acta Crystallogr. *B 30*, 1495 (1974)
Kobayashi, M.: Chem. Soc. Jpn. *64*, 648 (1939)
Kojima, M., Fujita, J.: Chem. Lett. *1970*, 429
Kojima, M., Yoshikawa, Y., Yamasaki, K.: Inorg. Nucl. Chem. Lett. *9*, 689 (1973)
Kojima, M., Iwagaki, M., Yoshikawa, Y., Fujita, J.: Bull. Chem. Soc. Jpn. *50*, 3216 (1977)
Konno, M., Marumo, F., Saito, Y.: Acta Crystallogr. *B 29*, 739 (1973)
Kuhn, W.: Z. Phys. Chem. *B 4*, 14 (1929)
Kuhn, W., Bein, K.: Z. Phys. Chem., Abt. *B 24*, 335 (1934a)
Kuhn, W., Bein, K.: Z. Anorg. Allg. Chem. *216* , 321 (1934b)
Kuhn, W.: Naturwissenschaften *19*, 289 (1938)
Kuroda, R., Saito, Y.: Acta Crystallogr. *B 30*, 2126 (1974)
Kuroda, R., Fujita, J., Saito, Y.: Chem. Lett. *1975*, 225
Kuroda, R., Shimanouchi, N., Saito, Y.: Acta Crystallogr. *B 31*, 931 (1975)
Kuroda, R., Saito, Y.: Bull. Chem. Soc. Jpn. *49*, 433 (1976)

Larsen, K. P., Toftlund, H.: Acta Chem. Scand. *A 31*, 182 (1977)
Larsen, S., Watson, K. J., Sargeson, A. M., Turnbull, K. R.: Chem. Commun. *1968*, 847
Legg, J. I., Douglas, B. E.: J. Amer. Chem. Soc. *88*, 2697 (1966)
Lennard-Jones, J. E., in: Fowler, R. H.: Statistical Mechanics, pp. 292–337, Camb. Univ.
    Press 1929
Liehr, A. D.: J. Phys. Chem. *68*, 665 (1964)
Lifson, S., Warshel, A.: J. Chem. Phys. *49*, 5116 (1968)
Lifson, S.: Molecular Forces. In: Protein-Protein Interactions. Jaenicke, R., Helmreich, E. (eds.).
    Berlin – Heidelberg – New York: Springer 1972
Lindoy, L. F., Busch, D. H.: Prep. Inorg. React. *6*, 1 (1971)
Liquori, A. M., Damiani, A., Elefante, G.: J. Mol. Biol. *33*, 439 (1968)
Liu, C. F., Ibers, J. A.: Inorg. Chem. *8*, 1911 (1969)
London, F.: Z. Physik. Chem. *B 11*, 222 (1930)
London, F.: Trans. Faraday Soc. *33*, 8 (1937)

MacDermott, T. E.: Inorg. Chim. Acta *2*, 81 (1968)
Manohar, H., Schwarzenbach, D.: Helv. Chim. Acta *57*, 1086 (1974)
Martin, R. B., Tsangaris, J. M., Chang, J. W.: J. Amer. Chem. Soc. *90*, 821 (1968)
Marumo, F., Utsumi, Y., Saito, Y.: Acta Crystallogr. *B 26*, 1492 (1970)
Marumo, F., Isobe, M., Saito, Y., Yagi, T., Akimoto, S.: Acta Crystallogr. *B 30*, 1904 (1974)

Marumo, F., Isobe, M., Akimoto, S.: Acta Crystallogr. *B 33,* 713 (1977)

Marzilli, P. A.: PhD. Thesis, Australian National University 1969

Mason, E. A., Kreevoy, M. M.: J. Amer. Chem. Soc. *77,* 5808 (1955)

Mason, S. F.: Quart. Rev. Chem. Soc. *17,* 20 (1963)

Mason, S. F., Norman, B. J.: Proc. Chem. Soc. *1964,* 339

Mason, S. F., Norman, B. J.: Chem. Commun. *1965,* 49

Mason, S. F.: Pure and Appl. Chem. *24,* 335 (1970)

Mason, S. F.: J. Chem. Soc. *A 1971,* 667

Mason, S. F., Seal, R. H.: Mol. Phys. *31,* 755 (1976)

Mason, S. F., Peart, B. J.: J. Chem. Soc., Dalton *1977,* 937

McMathieson, A. L.: Acta Crystallogr. *9,* 317 (1956)

Mathieu, J.-P.: J. Chim. Physicochim. Biol. *33,* 78 (1936)

Mathieu, J.-P.: Ann. Phys. *19,* 335 (1944)

Mathieu, J.-P.: Les Theories Moleculaires du Pouvoir Rotatoire Naturel. Paris: Gauthier-Villars 1946

Mathieus, D. A., Stucky, G. D.: J. Amer. Chem. Soc. *93,* 5954 (1971)

Matsumoto, K., Ooi, S., Kuroya, H.: Bull. Chem. Soc. Jpn. *43,* 3801 (1970)

Matsumoto, K., Ooi, S., Kuroya, H.: Bull. Chem. Soc. Jpn. *44,* 2721 (1971)

Matsumoto, K., Kuroya, H.: Bull. Chem. Soc. Jpn. *44,* 3491 (1971); *45,* 1755 (1972)

Matsumoto, K., Kawaguchi, H., Kuroya, H., Kawaguchi, S.: Bull. Chem. Soc. Jpn. *46,* 2424 (1973)

McCaffery, A. J., Mason, S. F.: Mol. Phys. *6,* 359 (1963)

McCaffery, A. J., Mason, S. F., Norman, B. J.: Proc. Chem. Soc. *1964,* 259

McCaffery, A. J., Mason, S. F., Ballard, R. E.: J. Chem. Soc. *1965,* 2883

McCaffery, A. J., Mason, S. F., Norman, B. J.: Chem. Commun. *1965a,* 29

McCaffery, A. J., Mason, S. F., Norman, B. J.: J. Chem. Soc. *1965b,* 5094

Meisenheimer, J., Angermann, L., Holsten, H.: Ann. *438,* 261 (1924)

Mikami, M., Kuroda, R., Konno, M., Saito, Y.: Acta Crystallogr. *B 33,* 1485 (1977)

Miyamae, H., Sato, S., Saito, Y.: Acta Crystallogr. *B 33,* 3391 (1977)

Miyamae, H., Sato, S., Saito, Y.: Acta Crystallogr., to be published (1979)

Mizukami, F., Ito, H., Fujita, J., Saito, K.: Bull. Chem. Jpn. *43,* 3973 (1970)

Mizukami, F., Ito, H., Fujita, J., Saito, K.: Bull. Chem. Soc. Jpn. *45,* 2129 (1972)

Moffitt, W.: J. Chem. Phys. *25,* 1189 (1956)

Morgan, G. T., Drew, H. D. K.: J. Chem. Soc. *1920,* 117, 1456

Morris, P. J., Martin, R. B.: Inorg. Chem. *10,* 964 (1971)

Muto, A., Marumo, F., Saito, Y.: Acta Crystallogr. *B 26,* 226 (1970)

Nagao, R., Marumo, F., Saito, Y.: Acta Crystallogr. *B 28,* 1852 (1972)

Nagao, R., Marumo, F., Saito, Y.: Acta Crystallogr. *B 29,* 2438 (1973)

Nakagawa, I., Shimanouchi, T.: Spectrochem. Acta *22,* 1707 (1966)

Nakahara, A., Saito, Y., Kuroya, H.: Bull. Chem. Soc. Jpn. *25,* 331 (1952)

Nakatsu, K.: Bull. Chem. Soc. Jpn. *30,* 795 (1962)

Nakayama, Y., Matsumoto, K., Ooi, S., Kuroya, H.: Chem. Commun. *1937,* 170

Niketić, S. R., Woldbye, F.: Acta Chem. Scand. *27,* 621 (1973a)

Niketić, S. R., Woldbye, F.: Acta Chem. Scand. *27,* 3811 (1973b)

Niketić, S. R.: Thesis, The Technical University of Denmark 1974

Niketić, S. R., Woldbye, F.: Acta Chem. Scand. *A 28,* 248 (1974)

Niketić, S. R., Rasmussen, Kj., Woldbye, F., Lifson, S.: Acta Chem. Scand. *A 30,* 485 (1976)

Niketić, S. R., Rasmussen, Kj.: The Consistent Force Field (Lecture Notes in Chemistry, Volume 3). Heidelberg: Springer 1977

Niketić, S. R., Rasmussen, Kj.: Acta Chem. Scand. *A 32,* 391 (1978)

Nishikawa, S., Matsukawa, K.: Proc. Imp. Acad. (Jpn.) *4,* 96 (1928)

Nomura, T., Marumo, F., Saito, Y.: Bull. Chem. Soc. Jpn. *42,* 1016 (1969)

Ogino, H., Takahashi, M., Tanaka, N.: Bull. Chem. Soc. Jpn. *43*, 424 (1970)
Ohba, S., Toriumi, K., Sato, S., Saito, Y.: Acta Crystallogr. *B 34*, 3535 (1978)
Okamoto, K., Tsukihara, T., Hidaka, J., Shimura, Y.: Chem. Lett. *1973*, 145
Okaya, Y., Saito, Y., Pepinsky, R.: Phys. Rev. *98*, 1857 (1955)
Ooi, S., Komiyama, Y., Saito, Y., Kuroya, H.: Bull. Chem. Soc. Jpn. *32*, 263 (1959)
Oonishi, I., Shibata, M., Marumo, F., Saito, Y.: Acta Crystallogr. *B 29*, 2448 (1973)
Oonishi, I., Sato, S., Saito, Y.: Acta Crystallogr. *B 31*, 1318 (1975)
Orgel, L. E.: Introduction to Transition-Metal Chemistry; Ligand Field Theory, London: Methuen 1960
Orville- Thomas, W. J.: Internal Rotation in Molecules. London: Wiley 1974

Pauling, L.: The Nature of the Chemical Bond, 3rd ed. Ithaca: Cornell Univ. Press 1960
Peterson, S. W., Smith, H. G.: J. Phys. Soc. Jpn. *17B-II*, 335 (1962)
Phillips, J. F., Royer, D. J.: Inorg. Chem. *4*, 616 (1965)
Piper, T. S., Karipides, A. G.: J. Amer. Chem. Soc. *86*, 5039 (1964)
Pitzer, K. S.: Adv. Chem. Phys. *2*, 59 (1959)
Poulet, H.: J. Chem. Phys. *59*, 584 (1962)

Ramachandran, G. N., Venkatachalam, C. M., Krimm, S.: Biophysical J. *6*, 849 (1966)
Rasmussen, Kj., Lifson, S.: Unpublished work; summarised in: Rasmussen, Kj.: Conformations and Vibrational Spectra of Tris-(diamine) Metal Complexes. Thesis, The Technical University of Denmark 1970
Raymond, K. N., Corfield, P. W. R., Ibers, J. A.: Inorg. Chem. *7*, 1362 (1968)
Raymond, K. N., Ibers, J. A.: Inorg. Chem. *7*, 2333 (1968)
Reihlen, H., Weinbrenner, E., v. Hessling, G.: Ann. *494*, 143 (1932)
Richardson, F. S.: J. Phys. Chem. *75*, 692 (1971a)
Richardson, F. S.: J. Chem. Phys. *54*, 2453 (1971b)
Richardson, F. S.: Inorg. Chem. *10*, 2121 (1971c)
Richardson, F. S.: J. Chem. Phys. *57*, 589 (1972a)
Richardson, F. S.: Inorg. Chem. *11*, 2366 (1972b)
Richardson, F. S., Caliga, D., Hilmes, G., Jenkins, J. J.: Mol. Phys. *30*, 257 (1975)
Robinson, W. T., Buckingham, D. A., Chandler, G., Marzilli, L. G., Sargeson, A. M.: Chem. Commun. *1969*, 539
Rosenfeld, L.: Z. Physik *52*, 161 (1928)

Saburi, M., Sawai, T., Yoshikawa, S.: Bull. Chem. Soc. Jpn. *45*, 1086 (1972)
Saburi, M., Sato, T., Yoshikawa, S.: Bull. Chem. Soc. Jpn. *49*, 2100 (1976)
Saito, Y., Nakatsu, K., Shiro, M., Kuroya, H.: Acta Crystallogr. *7*, 636 (1954); *8*, 729 (1955)
Saito, Y., Nakatsu, K., Shiro, M., Kuroya, H.: Bull. Chem. Soc. Jpn. *30*, 795 (1957)
Saito, Y., Iwasaki, H.: Bull. Chem. Soc. Jpn. *35*, 1131 (1962)
Saito, Y.: Coord. Chem. Revs. *13*, 305 (1974)
Sakurai, J. J.: Advanced Quantum Mechanics, pp. 20–68. Reading: Addison-Wesley 1967
Sakurai, T.: X-ray Crystal Structure Analysis, p. 219. Tokyo: Shokabo 1967 (in Japanese)
Sato, S., Saito, Y., Fujita, J., Ogino, H.: Inorg. Nucl. Chem. Letters *10*, 669 (1974)
Sato, S., Saito, Y.: Acta Crystallogr. *B 31*, 1378 (1975a)
Sato, S., Saito, Y.: Acta Crystallogr. *B 31*, 2456 (1975b)
Sato, S., Saito, Y.: Acta Crystallogr. *B 33*, 860 (1977)
Sato, S., Saito, Y.: Acta Crystallogr. *B 34*, 420 (1978)
Schäffer, C. E.: Proc. Roy. Soc. *A 1967*, 297; Pure and Appl. Chem. *24*, 361 (1970)
Schellman, J. A.: J. Chem. Phys. *44*, 55 (1966)
Schellman, J. A.: Accounts Chem. Res. *1*, 144 (1968)
Schellman, J. A.: Chem. Rev. *75*, 323 (1975)
Scherrer, P., Stoll, P.: Zeit. anorg. Chem. *121*, 319 (1922)

162   References

v. Schleyer, P. R.: in: Conformational Analysis, Scope and Present Limitations, p. 241, New York: Academic Press 1971
Schoenflies, A.: Krystallsysteme und Krystallstruktur, Leipzig 1891
Schwarzenbach, G., Moser, P.: Helv. Chim. Acta 36, 581 (1953)
Scouloudi, H., Carlisle, C. H.: Acta Crystallogr. 6, 651 (1963); Nature Lond. 166, 357 (1950)
Shinada, M.: J. Phys. Soc. Jpn. 19, 1607 (1964)
Shintani, H., Sato, S., Saito, Y.: Acta Crystallogr. B 31, 1981 (1975)
Shintani, H., Sato, S., Saito, Y.: Acta Crystallogr. B 32, 1184 (1976)
Shintani, H., Sato, S., Saito, Y.: Acta Crystallogr., to be published (1979)
Shull, C. G., Yamada, Y.: J. Phys. Soc. Jpn. 17, Supplement B-III, 1 (1962)
Snow, M. R.: J. Amer. Chem. Soc. 92, 3610 (1970)
Snow, M. R.: J. Chem. Soc. Dalton 1972, 1627
Stewart, R. F.: J. Chem. Phys. 53, 205 (1970)
Stoll, P.: Thesis, Zürich 1926
Strickland, R. W., Richardson, F. S.: Inorg. Chem. 12, 1025 (1973)
Sugano, S.: Chem. Phys. 33, 1883 (1960)

Tanaka, K., Marumo, F., Saito, Y.: Acta Crystallogr. B 29, 733 (1973)
Thewalt, U., Jensen, K. A., Schäffer, C. E.: Inorg. Chem. 11, 2129 (1972)
Tinoco, I.: Advan. Chem. Phys. 4, 67 (1962)
Toftlund, H., Pedersen, E.: Acta Chem. Scand. 26, 4019 (1971)
Toriumi, K., Saito, Y.: Acta Crystallogr. B 31, 1247 (1975)
Toriumi, K., Ozima, M., Akaogi, M., Saito, Y.: Acta Crystallogr. B 34, 1093 (1978)
Torrens, I. M.: Interatomic Potentials. New York: Academic Press 1972
Tsangaris, J., Martin, R. B.: J. Amer. Chem. Soc. 92, 4255 (1970)
Tsuchiya, H., Marumo, F., Saito, Y.: Acta Crystallogr. B 28, 1935 (1972)
Tsuchiya, H., Marumo, F., Saito, Y.: Acta Crystallogr. B 29, 659 (1973)

van Niekerk, J. N., Schoening, F. R. L.: Acta Crystallogr. 5, 196, 475, 499 (1952)

Watson, R. E., Freeman, A. J.: Phys. Rev. 120, 1134 (1960)
Wellman, K. M., Mecca, T. G., Mungall, M., Hare, C. R.: J. Amer. Chem. Soc. 90, 805 (1968)
Wellman, K. M., Bogdansky, S., Piontek, C., Hare, C. R., Mathieson, M.: Inorg. Chem. 8, 1025 (1969)
Wellman, K. M., Mungall, M., Mecca, T. G., Hare, C. R.: J. Amer. Chem. Soc. 89, 3647 (1969)
Werner, A.: Ber. 44, 1887 (1911)
Werner, A.: Ber. 45, 121 (1912a)
Werner, A.: Ber. 45, 1229 (1912b)
Whuler, A., Brouty, C., Spinat, P., Herpin, H.: Acta Crystallogr. B 31, 2069 (1975)
Wiberg, K. B.: J. Amer. Chem. Soc. 87, 1070 (1965)
Wiekliem, H. A., Hoard, J. L.: J. Amer. Chem. Soc. 81, 549 (1959)
Wigner, E. P.: Group Theory and its Application to the Quantum Mechanics of Atomic Spectra, p. 113. New York: Academic Press 1959
Williams, J. E., Stang, P. J., v. Schleyer, P. R.: Ann. Rev. Phys. Chem. 19, 531 (1968)
Williams, R. J., Larson, A. C., Cromer, D. T.: Acta Crystallogr. B 28, 858 (1972)
Wilson, Jr., E. W., Martin, R. B.: Inorg. Chem. 9, 528 (1970)
Wing, R. M., Eiss, R.: J. Amer. Chem. Soc. 92, 1929 (1970)
Witiak, D., Clardy, J. C., Martin, D. S.: Acta Crystallogr. B 28, 2694 (1972)
Witte, H., Wölfel, E.: Z. Physik. Chem. (Frankfurt) 3, 296 (1955)

Woldbye, F.: in: Techniques of Inorganic Chemistry. Jonassen, H. B., Weissberger, A.,(ed.), p. 249. New York: Wiley-Interscience 1965

Woldbye, F.: Proc. Roy. Soc. *A 297,* 79 (1967)

Woolfson, M. M.: An Introduction to X-Ray Crystallography, Cambridge 1970

Yaba, S., Yano, S., Yoshikawa, S.: Inorg. Nucl. Chem. Letters *12,* 831 (1976)

Yamatera, H.: Naturwissenschaften *44,* 375 (1957)

Yamatera, H.: Bull. Chem. Soc. Jpn. *31,* 95 (1958)

Yano, S., Furuhashi, K., Yoshikawa, S.: Bull. Chem. Soc. Jpn. *50,* 685 (1977)

Yoshikawa, S., Saburi, M., Sawai, T., Goto, M.: Proc. XII ICCC, Sydney, p. 155 (1969)

Yoshikawa, Y., Yamasaki, K.: Bull. Chem. Soc. Jpn. *46,* 3448 (1973)

Yoshikawa, Y.: Bull. Chem. Soc. Jpn. *49,* 159 (1976)

Subject Index

Inorganic Chemistry Concepts

Editors: M. Becke,
C. K. Jørgensen,
M. F. Lappert, S. J. Lippard,
J. L. Margrave, K. Niedenzu,
R. W. Parry, H. Yamatera

Inorganic chemistry has been stimulated by a variety of new developments in the past few years – by physics, due to the progress made in the use of lasers through the development of semiconductors and magnetic materials, and by crystal physics. Stimulation has come as well from the theoretical aspects of chemistry in the areas of stereochemistry, and in chemical structure and bonding theory.
The ensuing advances in inorganic chemistry will be the subject of this new series. The focus will be on fields of endeavor which as yet remain open and are therefore topical. The series is edited by an international panel of experts.

Springer-Verlag
Berlin
Heidelberg
New York